Cambridge Elements ☰

Elements in the Philosophy of Biology
edited by
Grant Ramsey
KU Leuven, Belgium
Michael Ruse
Florida State University

HUMAN NATURE

Grant Ramsey
KU Leuven, Belgium

CAMBRIDGE
UNIVERSITY PRESS

Shaftesbury Road, Cambridge CB2 8EA, United Kingdom

One Liberty Plaza, 20th Floor, New York, NY 10006, USA

477 Williamstown Road, Port Melbourne, VIC 3207, Australia

314–321, 3rd Floor, Plot 3, Splendor Forum, Jasola District Centre,
New Delhi – 110025, India

103 Penang Road, #05–06/07, Visioncrest Commercial, Singapore 238467

Cambridge University Press is part of Cambridge University Press & Assessment,
a department of the University of Cambridge.

We share the University's mission to contribute to society through the pursuit of
education, learning and research at the highest international levels of excellence.

www.cambridge.org
Information on this title: www.cambridge.org/9781009440646

DOI: 10.1017/9781108685481

First published 2023

A catalogue record for this publication is available from the British Library.

ISBN 978-1-009-44064-6 Hardback
ISBN 978-1-108-71606-2 Paperback
ISSN 2515-1126 (online)
ISSN 2515-1118 (print)

Human Nature

Elements in the Philosophy of Biology

DOI: 10.1017/9781108685481
First published online: July 2023

Grant Ramsey
KU Leuven, Belgium

Author for correspondence: Grant Ramsey, grant@theramseylab.org

Abstract: *Human nature* is frequently evoked to characterize our species and describe how it differs from others. But how should we understand this concept? What is the nature of a species? Some take our nature to be an essence and argue that because humans lack an essence, they also lack a nature. Others argue for nonessentialist ways of understanding human nature, which usually aim to provide criteria for sorting human traits into one of two bins, the one belonging to our nature and the other outside our nature. This Element argues that both the essentialist and trait bin approaches are misguided. Instead, the author develops a *trait cluster account* of human nature, which holds that human nature is based on the distribution of our traits over our (actual and possible) life histories. One benefit of this account is that it aligns human nature with the human sciences, rendering the central concern of the human sciences to be the study of human nature. This title is also available as Open Access on Cambridge Core.

Keywords: human nature, human uniqueness, essence, species, evolution

ISBNs: 9781009440646 (HB), 9781108716062 (PB), 9781108685481 (OC)
ISSNs: 2515-1126 (online), 2515-1118 (print)

Contents

1 Introduction

Humans. *Homo sapiens*. The upright ape once obscure and scattered like rare bits of stone across the broad expanse of Africa, now a world-dominating – a climatic, a geologic – force. What are we? What is our nature? We vary across cultures and history and we sort humans into a diverse array of categories. We can be scholarly, sleazy, shallow, sassy, saturnine. We vary in our tastes, abilities, and dispositions to such a degree that true universal generalizations about humans, at least ones not disjunctive or vague or trivial, can be difficult to find.

What unifies Yaminawa living in the remote Peruvian Amazon with Buddhist monks in Thailand with Wall Street traders with Haitian vodou priests with California yoga instructors? What these individuals take as important features of the world – what they even take to be real features of the world – are radically divergent. Does human nature lie in what is the same across such diverse sets of people? Or does looking only for commonalities eliminate most of what is interesting and important about our nature?

If we ask, What is our nature?, there are two quite different sets of questions we may have in mind. One set concerns our character. Are we humans good, though subject to corruption? Or are we evil, possessing a wickedness that can perhaps be tamed, repressed, or obliterated? Questions of this kind take for granted that generalizations like "we are good by nature" are not problematic, that we have a nature and our task is simply to determine its features.

But there are prior, more basic questions. If I am describing the color of things, I may characterize them as azure, crimson, mauve, or taupe. But I could also step back and ask: What is color? What does it mean to say that something is colored? Is the color we see an intrinsic property of objects or an artifact of human visual systems? For human nature, the analogous prior questions are: What does it mean for a species to have a nature? What does the phrase "human nature" refer to? As we will see in the following section, some researchers have expressed skepticism about human nature, at times even suggesting that humans lack a nature, or that the very idea of a species nature is problematic.

There is thus a puzzle to solve prior to elaborating our nature. We must first reflect on the very concept of a species nature. Only then can we determine whether talk of human nature is justified. And if it is, we can push forward to develop a coherent – and perhaps even useful – concept of human nature.

This Element centers on these more fundamental questions. It is not concerned with how we can or should record, describe, and understand our features. It is not a methodological treatise. Nor is it an attempt to offer enlightening generalizations about our species (as selfish or altruistic or such). It is not an empirical investigation into how and why we behave in the way that we do. And it sits outside of *philosophical anthropology* as traditionally practiced, which, in the words of Peter Hacker, "is the study of the conceptual framework in terms of which we think about, speak about, and investigate man (*Homo sapiens*) as a social and cultural animal" (2021, xi). Finally, it is not a survey of what are sometimes called "theories of human nature," a rubric under which falls a collection of Western and Eastern scientific, philosophical, and religious views on the world and ourselves – such as Buddhism, Confucianism, existentialism, feminism, Freudianism, Marxism, and Platonism (Stevenson et al. 2017).

Instead, the focus is simply on what we mean – what we should mean – by *human nature* within a contemporary scientific worldview. When we say of a behavior that it is natural, what does this amount to? What empirical claim is being advanced? If we say of a trait that it is *part of our nature*, what might this mean? What concept of nature is at play in such a statement? What good (if any) does, or can, the concept of human nature play in the sciences? How should we understand popular discourse, in newspapers and magazines and social media, citing human nature as a cause of our actions? How has human nature been defined and critiqued, and which human nature concept might be the best for fulfilling the roles that we desire it to play?

We begin this journey in Section 2 by considering and rejecting human nature skepticism, the position that humans do not, in fact, possess a nature. I conclude that if we hold that a nature is an essence, then human nature is rightly rejected. But if we interpret human nature in a nonessentialist manner, we can sidestep some of the critiques waged by human nature skeptics.

The next task, in Section 3, is to evaluate the main way that human nature has been conceptualized, what I label the "trait bin" approach. The trait bin approach to human nature holds that the key to divining human nature is to sort traits into one of two bins, the human nature bin and the remainder bin. I argue that this approach, though it has a strong initial appeal due to its simplicity, is ultimately untenable.

From there, I move on in Section 4 to develop an alternative to the trait bin approach, which I label the "trait cluster" approach. The trait cluster approach centers on the idea that our nature is not defined by a bin of traits shared by all or most humans, but instead lies in how traits are exhibited within and across human life histories.

This trait cluster account has some counterintuitive implications and is easily misunderstood. Thus, in Section 5, I examine some critiques of the view. One critique is that my account is too permissive, and that this permissiveness makes it vacuous. I argue that the permissiveness critiques arise from a failure to fully grasp the difference between a trait bin and trait cluster account. Another critique relates to the core–veneer distinction. Human nature is often thought of as residing in our core, implying that we have a natural core overlain by a cultural veneer. I argue against the view that human nature is about what's within, about our core. Instead, I conclude that the core–veneer distinction has fundamental problems and, furthermore, is unnecessary to understanding ourselves.

Having elaborated and defended the trait cluster concept of human nature, the task of Section 6, the last one before the conclusion, is to explore what the trait cluster account of human nature can do. Can it explain occurrences of traits? Can it be used to learn about our nature and how it differs from the nature of other species through the study of twins, triplets, and even related species, such as chimpanzees? Can it make sense of how the concept of human nature is used in popular media? Can it be a guide to our moral behavior?

Let's now begin our journey through these topics, starting with human nature skepticism.

2 Against Human Nature

An argument for the hopelessness of seeking a coherent, productive, useful concept of human nature was offered by David Hull in his article "On Human Nature" (1986). This article has cast a long shadow over the field, and it is thus important to understand what his argument is and whether it is convincing. To grasp his argument, we first need to understand what he takes a human to be. Only then can we follow how he derives his argument that humans lack a nature.

Before we begin, however, it is worth pointing out how radical the idea is that there is no such thing as human nature. As David Hume argued in his monumental *A Treatise of Human Nature*:

> 'Tis evident, that all the sciences have a relation, greater or less, to human nature; and that however wide any of them may seem to run from it, they still return back by one passage or another. Even Mathematics, Natural Philosophy, and Natural Religion, are in some measure dependent on the science of Man; since they lie under the cognizance of men, and are judged of by their powers and faculties. (1731, xix)

He thus placed human nature at the very foundation of human inquiry and linked human nature with the "science of Man." Could it be that Hume's seven-hundred-page treatise on human nature is not on anything at all? Was he chasing a mirage for hundreds of pages? Hume was not alone in positing the centrality of human nature. Thus, to argue that there is no such thing as human nature calls into question broad swaths of philosophy. It is therefore radical and highly destructive – and should not be accepted without an airtight argument to back it up.

2.1 What Is a Human?

What is a human? There are two basic responses we may offer. The simplest answer is that an organism is a human if and only if it is a member of the scientific category *Homo sapiens*. This is the response given by Hull (1986). But before we consider this response in detail, we will first consider a second response, which relies not on a scientific criterion, but on normative evaluations – that is, evaluations of what we *ought* to be like, not merely what we *are* like. It is thus prescriptive, not merely descriptive. If we talk of certain behaviors as being "inhuman" or of certain people being or acting (merely) like animals, then we are using normative criteria. One can be inhuman in the normative sense while still belonging to *Homo sapiens*.

This normative way of defining humans has clear problems. One is that it renders human nature (at least in part) merely stipulative. If we add our own normative criteria for what it is to be a proper human (to act morally, say) to the concept of a human, then it is not an empirical question whether we are moral. Humans will be moral because we stipulate that humans are moral creatures. But if we maintain that whether a species (*Homo sapiens* or otherwise) exhibits a particular trait (like morality) should be an empirical matter, then trait possession should be discovered, not stipulated. We therefore need to avoid using a normative mold to cast the boundaries of species. It is for this reason preferable to take humans to simply be members of *Homo sapiens*. While being moral may be an important feature of our kind, it is not the basis of our species's boundary.

You may respond by asserting that normativity must be a part of what it is to be a human, since human nature is deeply normative: human nature is about how we should be, not about how we are. One motivation for this position is that it is appealing to have human nature be a target at which to aim. If human nature is an ideal target, then deviations from it can be reasons to strive to be more like it. We may even hold that we *ought* to strive in this way.

But if human nature is an ideal of this kind, it is an invention – a product of culture and imagination. Such ideals often crystallize in religion, where disciples are told how they ought to be, what it is to be a good Christian or Muslim or Hindu. The ideals offered by religions can be deeply meaningful belief systems that shape human behavior and profoundly inform our understanding of ourselves and others. In this way, they provide insight into our nature: we are a species capable of generating complex, meaningful religious systems, and this is a fact important to understanding ourselves. But acknowledging this does not mean that particular religious ideals are themselves true accounts of our nature. For example, the Catholic Church, as they describe in their *Catechism* (part one, section two, chapter one, article one, paragraph seven[1]), maintains that we should interpret the biblical Adam and Eve story in the following way:

> 417 Adam and Eve transmitted to their descendants human nature wounded by their own first sin and hence deprived of original holiness and justice; this deprivation is called "original sin."
> 418 As a result of original sin, human nature is weakened in its powers, subject to ignorance, suffering and the domination of death, and inclined to sin (this inclination is called "concupiscence").

The lesson to take from this is that we are a species capable of generating and believing these kinds of stories. We should not, however, conclude that we have a weakened nature due to original sin. Instead, this tale can be appreciated as one among countless tales about our origin and nature generated by a diversity of religions the world over (Ramsey 2017).

We are a culturally diverse species, and there are many standards offered by cultures and religions for how to behave, for how to be a proper human. But we need to distinguish *what we are like* from *what we think we should be like*. As I will argue below, if we want human nature to align with the sciences, then it is the former that is human nature. The latter can – depending on how, precisely, human nature is understood – be seen as an aspect of our nature or as partially due to our nature. Thus, while the moral dimensions and implications of human nature are important, they will enter this Element only after we figure out what human nature is.

Thus, eschewing normative ways of defining humans, "human" will simply denote *Homo sapiens*. A member of this species is a human no matter how they act. The least moral human is still a human. Of course, this immediately prompts us to ask what the criteria are that make each of us belong to the biological species *Homo sapiens*. The answer to this comes from knowing what it is for something to belong to a species. And to know this, we must have a strong grasp on the Darwinian insight that the history of life on Earth has a tree structure, where species are branches on this tree – an insight that played a crucial role in Hull's argument.

2.2 Essences and Evolutionary Trees

Prior to Charles Darwin's publication of *On the Origin of Species* in 1859, species were often thought of as having an immutable essence. A dog was in essence a dog, and its offspring were dogs because they inherited this essence. One could breed a dog and achieve individuals as diverse as Irish wolfhounds

---

[1] www.vatican.va/archive/ENG0015/__P1C.HTM.

and Chihuahuas. But these are all dogs, and while you can breed them to be different in size, color, and such, you cannot breed a dog into a cat or a horse. Cats, dogs, horses, and all other species are essentially different. Literally, that is: they contain distinct essences. Under one interpretation, these essences were divinely created. Dogs were created by God as dogs, and forever they will remain so.

The view of species as fixed types with unique essences underwent a powerful critique by Darwin. In the *Origin*, he did two important things. First, he argued that natural selection is a powerful and creative force, a force capable of generating adaptations – traits that fit their function. Eyes well adapted for seeing food, prey, and predators, and teeth sharp for piercing flesh or flat for grinding grass or seeds.

Second, he argued that the history of life has a tree structure. That is, not only do species have an evolutionary history, but they also have common ancestors: trace any two species back in time far enough and you come to a point when they were one. We now know that humans and chimpanzees arose from the same species more than six million years ago, humans and orangutans close to thirteen million years ago (Glazko and Nei 2003; Almécija et al. 2021).

Darwin's tree of life view is a radical departure from accounts that take species to be independently created. It gave new meaning to the shared traits among creatures. The bones in a bat's wing and our hand are the same not because of some shared divine blueprint, but because we share an ancestor with the same bone structure, and bone correspondences (though not necessarily shape) are well conserved over time. This correspondence relation is that of *homology*, a concept introduced in a non-evolutionary framework by Owen (1843) but reinterpreted by Darwin (1859) to be based on shared ancestry (Ramsey and Peterson 2012).

Darwin's view also gave new meaning to the species concept: extant species are just terminal branches on the tree of life. What is crucial to understand about the tree structure of the history of life is what it implies for the answer to this question: *Why does a given organism belong to a given species?* The Darwinian answer, which is the answer that contemporary evolutionary biology also provides, is that the organism belongs to the species not because of an intrinsic property it possesses, but merely because of where the organism is located on the tree of life. If the organism is within the branch of the tree constituting the species, then it belongs to that species even if it deviates from the norm.

This is not entirely true, since some deviations from the norm can be extreme enough that new species are founded. It is thus not the case that an organism belongs to a given species if and only if the organism's parents belong to that species, since this would make speciation impossible. On the contrary, branches can split and new species can form. Speciation is a complicated process often involving periods of ambiguity. The complications of speciation will not be dealt with here, since all we need is the understanding that a species is a branch and that belonging to a species is belonging to its branch. It is also important to mention that while these relational (instead of intrinsic) properties are what makes an individual a member of a species, this does not mean that intrinsic properties are not important to species determinations. On the contrary, intrinsic properties play important evidentiary roles in classification. If it looks like a duck and quacks like a duck, then probably it is a duck. The quacking and appearance can thus provide evidence concerning its classification.

We can visualize Darwin's framework as a tree laden with fruit. It is an unusual tree, with different branches grafted on from different fruit varieties. One branch produces apples, another pears, yet another cherries. Now consider a single fruit. What makes it a pear? It is tempting to point to the fruit's bottom-heavy shape, the unique floral taste, the waxy yellowish-green maculated skin. This answer points to the fruit's intrinsic properties, and these properties may be very useful in identifying pears, but it is important not to mistake the usefulness of these properties with what makes something a pear. In this case, it is a pear because it is on the pear branch. A small round fruit growing on this branch will still be a pear, though a strange one. An apple that looks and tastes much like a pear will still be an apple so long as it is on the apple branch. The properties are merely useful guides, not necessary and sufficient conditions for membership in their kind.

We therefore need to separate the main distinguishing characteristics of an organism from what makes the organism belong to its species. We may think of the Asian one-horned rhinoceros as being large, gray, having one horn, herbivorous, and so on. And using these criteria may work flawlessly in picking out *Rhinoceros unicornis* from a lineup of mammals. But it is not these properties that make it belong to *R. unicornis*. Instead, it is its location in the tree of life. If an *R. unicornis* mother gives birth to a hornless offspring, it will still be a rhino, despite lacking a key distinguishing feature. (Again, I don't wish to imply that speciation is impossible. It is improbable though certainly possible that the hornless rhino marks the saltational beginning of a new hornless rhino species. But if it is a founding member of a species, this is the case because of the branch it forms, not merely because it substantially differs from its parents. The point is that in the absence of the founding of a new species-level branch, the individual is in the same species as its parents, despite its aberrant traits.)

What are the implications of the Darwinian view of species for human nature, and for Hull's skepticism about human nature? If "human" means *Homo sapiens*, and if belonging to *H. sapiens* is based not on attributes we think of as important human traits – speaking language, being moral – then these traits are not definitive of our species. They are not our essence. They are common, but not necessary to being human. Each of us is human based on our place in the tree of life, not on our characteristics. Diverse ways of being human do not make us more or less human. Someone who is bisexual, lesbian, asexual, or transgender is human independent of sexual preference or gender identity, no matter how unusual.

2.3 Hull's Argument

With the nonessentialist concept of human in hand, we are almost ready to lay out Hull's argument. But before we get there, we must understand his concept of a *nature*. For Hull, a nature is an essence – it refers to the necessary and sufficient conditions for membership in a kind. While there may be such things as relational essences, the essences Hull is concerned with involve intrinsic properties.

We can now see how Hull's argument gets off the ground. He combines a nonessentialist notion of a species with a nature understood as an intrinsic essence. Doing so appears to problematically refer to the essence of a nonessentialist collection of things. If humans have no essence, it seems to follow that they have no nature. And if this is true, human nature is an incoherent concept – it is an incompatible marriage of an essence-free scientific category (species) with an essence. We could summarize Hull's argument as follows:

Premise 1: The human species has no essence.

Premise 2: Natures are essences.

Therefore: The human species has no nature.

There is a variety of responses we can have to this argument. If we take it to be sound, it appears that we should stop talking about human nature. In this view, human nature is like the élan vital, the vital force that living things possess but the nonliving lack. The élan vital, once taken seriously by biologists, has since been discarded. There is no place for it in the contemporary metaphysics of biology. This is how Hull responds to this argument. Human nature, insofar as it is understood to rest on a scientific foundation, is a mirage. It may appear to exist when viewed at a distance, but on closer inspection, it is absent.

Another response to Hull's argument is to challenge its soundness by challenging the truth of one or more of its premises. Premise 1 is not readily challenged since, as we saw, it appears to follow from the structure of the tree of life. I should note, however, that some argue for forms of essentialism that could apply to Darwinian species, such as *historical essentialism* or *origin essentialism* (Griffiths 1999; LaPorte 2004). You might think that having the parents you do is a part of your essence, or that having a particular ancestor is part of the essence of an individual organism. There are debates over whether evolutionary trees support historical essences (Pedroso 2012), but such arguments are orthogonal to the point here. Hull was concerned with intrinsic property essences, so pointing out that you can get a kind of essence "for free," since individuals essentially have the ancestors they do, does not bear on Hull's argument. I won't further discuss historical or origin or other relational forms of essentialism in this Element.

Let us therefore assume that Premise 1 is true and instead turn to Premise 2, which equates essences and natures. I agree with Hull that if natures are essences, there is no such thing as human nature. When the biologist Michael Ghiselin wrote, "What does evolution teach us about human nature? It tells us that human nature is a superstition" (1997, 1), he likely had the equation of nature and essence in mind. It would a superstition be if we held this equation.

But while the terms "essence" and "nature" are often considered synonymous, they need not be. While it is clear that Hull's challenge is an important one – we must concede that if human nature picks out the essence tying us to our species, then we run into problems – it is possible to reject the second premise and assert that human nature is not essential to our belonging to *Homo sapiens*.

In fact, when the term "nature" is used, it often refers not to an essence, but instead picks out the important features of something. If you say that a lion is aggressive by nature, you are probably not meaning that there is some hidden essence within the lion associated with these behaviors. Nor are you claiming that being aggressive is a necessary property for being a lion, such that no nonaggressive animals can be lions. Instead, you presumably mean that lions are disposed toward aggressive behavior. Such a disposition is grounded in the psychology and physiology of the organism. In this view, generalizing about the species amounts to saying that the disposition is, at minimum, widespread.

A nature in this sense is like a family resemblance. Intelligence and petiteness might run in your family. You could thus rightly characterize your family as smart and petite. But this in not incompatible with a dull or stout individual being born into your family. These properties are not essences and therefore do not mean that the person is not, in fact, a part of your family.

Thus, a nature could be linked to dispositions to express traits, or to the pattern of trait expression, or even to a mere subset of human traits. If a trait is only widespread and not universal in a species, then it cannot be definitive of that species. A nature linked to such traits is thus not an essence. If such nonessentialist conceptions are allowed, Premise 2 is false. Natures are not (in this context, at least) essences.

There thus appear to be two options for how to react to Hull's argument: (1) take human nature to be an essence and endorse his conclusion that there is no such thing as human nature, or (2) explore nonessentialist concepts of human nature. One reason to pursue (2) instead of (1) is the entrenched usage of the term "human nature." Entire books are published on the topic. In *The Blank Slate: The Modern Denial of Human Nature*, psychologist Steven Pinker spent over 500 pages defending the idea that humans have a nature (Pinker 2002). (This is not to say that such books are unproblematic; Pinker's has been criticized for, among other things, an overly simplistic characterization of the distinction between blank-slate proponents and biological determinists.)

The use of the human nature concept extends far beyond philosophical and popular science treatments. It appears in the news media playing roles in making sense of and explaining our behavior. Consider a few examples from the *New York Times* with "human nature" in the title: John R. Quain (2016), "Makers of Self-Driving Cars Ask What to Do with Human Nature"; Evan Lipkis (2017), "Blame Human Nature, Not Guns"; and Farhad Manjoo (2018), "The Problem with Fixing WhatsApp? Human Nature Might Get in the Way." Such articles assume that there is such a thing called human nature, and they take it to have explanatory force: we can cite our nature in explaining our behavior. Are these authors deeply confused about human nature? Assuming that they are not referring to essences, then what is it that can do the explaining? (I will return to these examples in Section 6.4 to see whether the way they employ the concept of human nature can be understood in terms of the framework argued for below.)

Given the entrenched state of the discourse surrounding human nature, this question is worth answering. Instead of simply declaring that the notion of human nature is incoherent, we should pause and ask what it is that people like Pinker mean when they defend the idea that there is a human nature. Instead of trying to suppress talk of human nature, it thus may be more fruitful to explore and explicate what human nature might mean. This Element centers on the development of a nonessentialist conception of human nature that can help us to reject Hull's argument and make sense of what we mean when we are talking about human nature.

Simply rejecting the equation of intrinsic essences and natures in the case of human nature does not thereby provide us with a nonessentialist concept of human nature. Instead, we must develop one. The development of such a concept will be the chief focus of the next two sections. In them I will consider and reject a tempting nonessentialist conception of human nature (Section 3) before developing my own account (Section 4).

3 Is Human Nature a Bin of Traits?

The conclusion of the previous section is that if we have any chance of producing a concept of human nature that is coherent, that doesn't contravene known science, that can serve the roles desired of it in scientific and popular discourse, it cannot be based on intrinsic essences. But to say that human nature must be nonessentialist doesn't get us very far. What should such an account look like? What should it be built out of?

Is it human nature to be selfish? Are humans by nature altruistic? Such questions concern behaviors and psychological states – their disposition to be expressed and the motivations behind them – in this case selfish or generous ones. These are a kind of phenotypic trait, though not the only kind. Biologists typically distinguish *behavioral traits* from *morphological traits* (from the Greek *morpho*, meaning shape or form, these are the structural traits). Both kinds of traits are linked to human nature. The complete set of traits is highly diverse. Some traits are fleeting, others enduring. Height and weight are traits that change only a small amount from day to day. At the other end of the scale are ephemeral behaviors, such as the blink of an eye or flick of the finger. Some are intentional (you beating a drum) and others outside your direct control (your beating heart). Some are performative (singing a song) and others psychological (imagining yourself singing). But all are traits, and it is traits that form the basic ingredients of human nature.

Knowing the ingredients is not enough, however. You won't get cookies unless you know how to combine the ingredients and divide and bake the dough. Similarly, we need to know how traits get combined and baked into human nature.

The most straightforward way of relating human nature and human traits is to hold that traits can be divided into two sorts, human nature traits and traits outside of our nature. This is the relation philosopher David Buller (2005) has in mind when he observes that "the concept of human nature has traditionally designated only a proper subset of human behavior and mentation, which was claimed to belong to human beings *by their* nature as opposed to behavior and mentation that was claimed not to be owing to or in accordance with that nature" (420). Accounts that hold that human nature should be considered a subset – a *bin* – of traits will be labeled *trait bin accounts*. There are many ways that one could specify a trait bin account. In this section, I will begin by considering philosopher Edouard Machery's (2008) "nomological account." This is a good test case since it is clearly defined and explicitly defended.

3.1 Machery's Trait Bin Account

Machery defines the human nature bin in the following way: "Human nature is the set of properties that humans tend to possess as a result of the evolution of their species" (2008, 323). By *properties* he means traits in the sense described above. By *tend* he refers to a statistical quantity, which he specifies elsewhere as a majority, that is, more than 50 percent. If the trait is in the human nature bin, at least 50 percent of humans must possess it.

Trait bin accounts thus take the entire set of human traits and segregate a subset of them into the human nature bin. All trait bin accounts are the same in this way. How they differ is in the properties they use for segregation. Machery uses two individually necessary and jointly sufficient properties. One is a statistical property, that is, not an intrinsic property of the trait, but merely its statistical frequency. He adds to this a historical property, that the trait must have a certain kind of history. Specifically, the trait must be a result of the evolution of the species in question, in this case, *Homo sapiens*.

I will first critique Machery's segregation criteria before issuing a more general critique of the trait bin approach to human nature. To begin, let us trace out the implications of Machery's criteria. His first criterion – that the trait must be possessed by the majority of our species – is motivated by the idea that we should be able to take our characterization of human nature and use it in the manner of a field guide. The descriptions and illustrations in field guides attempt to characterize typical members of the species through universal (or at least common) traits, and to distinguish one species from another. Thus, if we have a lineup of primates, a field guide characterization of human nature would allow us to easily pick out the humans from the lineup. The less common the trait, the less useful it will be for that purpose.

But here we should draw a distinction between being useful for *identifying* humans and being useful for *characterizing* our species. Humans are plenty easy to identify, and we don't need a concept of human nature to try to distinguish members of our species from related taxa. This is in contrast to birds or beetles, in which the richness of species and the similarity of congeners can make species identification a challenge. If human nature is at all useful, it is in characterizing our species, not in giving us tools for picking out *Homo sapiens* individuals in the wild. If the genus *Homo* had a dozen extant species, things would be quite different. In such a world, a guide for distinguishing the various *Homo* species would be welcome.

If the goal of human nature is to characterize our species instead of serving as a species identification tool, then some of the traits important for this characterization will be ones exhibited by a minority of our species. A minority of humans undergo menopause, but this is an interesting trait, one quite unusual among mammals and in need of explanation. In fact, we appear to be the sole terrestrial mammal exhibiting the trait. The only other taxa known to undergo menopause are toothed whales (Ellis et al. 2018). To say that menopause is not part of our nature because the majority of humans does not experience it seems arbitrary and unhelpful.

In addition to sexually dimorphic traits (traits exhibited by only one sex), there are many traits that are distinctive of a species, but not exhibited by a majority. Take the leaf cutter ant species *Atta cephalotes*. Each colony has a queen, who is large, long-lived, and fertile. She is the only sexually reproductive female in the group and is the mother of the millions of individuals that compose a mature colony. Other casts include the soldiers, who have fierce mandibles and will pour forth from the colony when disturbed. Then there are the workers, who venture out to harvest leaves for the colony's subterranean fungus gardens. Finally, there are the minims, tiny workers specialized in tending the gardens. Some of these minims hitchhike on the leaves and watch for phorid flies, parasites that lay eggs on the leaves, which get carried into the nest to hatch and feed on the ant larvae. Finally, there are the males, which are winged and short-lived.

What is the nature of *Atta cephalotes*? To insist that we should include only traits that the majority of members of the species possess would leave out the traits of males, soldiers, minims, and queens. We would be left with an incredibly poor understanding of the species. The same is true of humans: insisting that only majority traits count as part of our nature will leave out many of our most interesting and distinctive features.

The majority criterion, while having a well-intended motivation, is thus not an effective way to delimit human nature. On top of the fact that it does a poor job in characterizing our species, it also excludes much of human diversity from human nature. If a 50 percent criterion is problematic, it may be thought that the easiest solution is to adjust the percentage. A more inclusive 40 percent cutoff – that is, traits must exist merely in at least 40 percent of the population to be considered a part of our nature – would capture some traits not previously included, but ultimately any criterion will be arbitrary and problematic. In fact, the basic idea of needing a statistical cutoff criterion is suspect. More will be said on this point below. But before we get there, let's consider Machery's second criterion, that human nature traits must be a product of the evolution of our species.

The evolution criterion is meant to distinguish traits due to our nature from those "exclusively due to enculturation or to social learning" (Machery 2008, 326). To see what is wrong here requires reflecting on the nature of development. All traits are a result of complex sets of causes. These causes include environmental inputs as well as inherited elements – principally genes and epigenetic factors. Let's focus on genes. They play a role through coding proteins and by regulating how other genes behave. Genes are inherited, and the set of genes you have is the result of a long evolutionary history. What follows from this is that any trait you have is built out of this history and is thus due to evolution in this sense.

You may object to this by saying that surely some traits are due only to social learning. Reading, break dancing, and gambling – all cultural traits for that matter – are not in our genes. They are due only to social learning, no? The answer to this is subtle, since it requires sharply distinguishing *what causes a trait* in a single individual (you reading Nietzsche, say) from *what causes a difference in traits* across two or more individuals (you reading Nietzsche but your friend reading Kant).

Consider two humans, one with blue eyes and one with brown. What lies at the basis of this difference? The difference is due to a genetic variation, a mutation in the OCA2 gene, which most likely arose 6,000–10,000 years ago (Eiberg et al. 2008). How about singing the blues? There is no gene that can be identified as having arisen in the past that underlies the difference between singing the blues and not. This is a cultural trait par excellence. Culture, and the social learning of cultural variants from others, is what explains why one individual sings the blues and another doesn't. (General singing abilities may have undergone selection, and there is, of course, variation in the ability to sing, so this example is best understood as asking why, among individuals who are able to sing, some sing the type of music they do instead of another.) But this is about what causes a *difference* in traits, not in what leads a particular individual to sing the blues. Surely genes, and by extension their evolutionary history, led to this singing behavior in individuals. One way of seeing this is to observe that, while for two individuals the difference between singing and not singing the blues may be cultural, this is not true for other pairs of individuals. The difference between a blues-singing human and a non-blues-singing chimpanzee is not a cultural one. No exposure to blues culture will get a chimp to be a Muddy Waters or a B. B. King.

The lesson to draw from this is that whether it is genes or culture that explains trait differences among individuals depends on the individuals in question. Furthermore, the fact that the difference is cultural only for a particular pair does not imply that in their individual development of the trait, genes played no role. This is why we need to be careful in how we interpret the concept of *heritability*. The heritability of a phenotypic trait is the measure of the degree to which variation in the trait in a population is due to variation in the genes of the members of that population. It is tempting to infer from a high heritability value to the conclusion that genes – or in particular, the genes underlying the variation in question – play a special role in the individual production of that trait. But this is not a valid inference.

That it is invalid is best seen through the fact that heritability values can be changed without changing genes. Take a population of fir trees in a highly heterogeneous habitat. Some patches of ground are very rocky and resource poor while others are rich in organic material and nutrients. Some of the fir trees are in the shade of hemlocks or spruces, while others are in full sun. Now take a phenotypic trait – height at five years, say – and measure its heritability. In a very heterogeneous environment such as this, heritability will be low. That is, what best explains differences in tree height are the environmental variables. But now imagine if the exact same genetic stock were in a very homogeneous environment, one in which every tree finds itself in rich soil and full sun. The heritability in such an environment will be high since genetic variation will play a much stronger role in the differences in height.

It should now be clear that trying to sort traits into those due to evolution and those due to the environment (including culture) is misguided. Is the height of a tree due to the evolution of its species or the environmental conditions it finds itself in? This question is clearly flawed. If we take a pair of trees, we may ask what explains their difference in height. In some cases, it may be due only to genes or only to the environment. But the fact that this is true for some pairs of trees does not mean that, within individual trees, genes or the environment plays a singular role in height determination. For individuals, the roles of genes and environment are inextricably intertwined.

A tempting reply to this argument is to point to knockout experiments as a way of showing whether a trait is due to a gene. A knockout experiment occurs when a specific gene is deactivated. If we compare a normal mouse with one that has a gene knocked out, we can, it seems, see what role that gene plays in the formation of traits. But here, again, we have the problem of taking differences among individuals and trying to make inferences about the role of the gene within individuals. If the knockout mouse has white fur instead of brown fur, it is tempting to claim that the gene is "for" brown fur. But in fact all we know is that for these two mice, this genetic difference underlies this phenotypic differ-ence. Genes play roles in networks, and the network was disrupted by the knockout. We do not know whether this gene evolved for making the brown color, or even if it has a privileged role in making it.

Compare this to a case of vandalism in a car lot. Half the cars will not start, and you find that their fuel lines have been cut. It may be tempting to infer that the fuel line is "for" starting the car, but, of course, there are many ways for the car to fail to start: knock out the starter motor or battery or spark plugs, for example. Any of these will keep the car from starting. Thus, it would be a mistake to point to just one as *the cause* of the car starting. The same is true of the knockout mouse and the same is true of any trait. We need to be extremely cautious in taking differences (car not starting versus car starting, brown fur versus white fur) and making inferences about the special role of whatever happens to underlie the difference.

On top of the problem of segregating traits into those exclusively due to genes or "exclusively due to enculturation or to social learning" (Machery 2008, 326), there is another major problem with the trait bin account: What do we do with quantitative traits? Some traits are qualitative – you either have them or you don't. I have a stomach and eyes, and you probability do as well. But many traits are quantitative. It is not that I either have height or I do not, but that I have a particular height, in my case 193 cm.

Each instance of a quantitative trait (my weight today, for example) will have a particular value. Measured precisely enough, each person has a unique height and weight. If a trait has to occur in more than half of the species for it to count as human nature, how should we understand quantitative traits like height and weight? Since each person has a unique height, is one's height never a feature of human nature? Or should we come up with height categories, such as 150–200 cm, and tally the number of individuals in each category?

Any such category system would be arbitrary, and we could choose a coarse-grained system that makes one category a part of our nature, while the others are outside our nature. Or we could implement a fine-grained categorization in which no height category is part of our nature, since no category contains at least 50 percent of our species. It is hard to justify either approach, and it is not clear what insight either brings to the understanding of our nature.

Let's pause and take stock. Human nature is frequently understood in what I am labeling a *trait bin approach*. Such an approach requires finding criteria for segregating traits that are a part of our nature from those outside of our nature. The example I used was that of Machery (2008), who proposed that two criteria should be used, the *evolution* and *frequency* of the trait. These criteria were shown to be problematic, but the critique of these criteria exposed some deeper problems with the trait bin approach. One is the problem of quantitative traits, which are not apt for being sorted into discrete categories, at least not without first converting quantitative traits into qualitative traits by drawing boundaries. But if we draw these boundaries, the traits that end up in human nature will be there simply because of how we draw the boundaries, not because of the intrinsic importance of the traits, and not because of the causal processes that bring about the traits in individuals.

Furthermore, while the evolution criterion seems like a wonderfully objective criterion for sorting traits into bins, it breaks down when we try to drive a wedge between traits due to evolution and those not due to evolution. In fact, any attempt to carve off culture to see the nature beneath will fail. Again, culture and genes can explain trait differences across individuals, but this does not mean that they play a privileged role in the development of the trait within those individuals.

Is there any hope of saving the trait bin approach? A viable trait bin approach would have to offer a way of converting quantitative traits into qualitative traits and criteria for sorting these newly minted qualitative traits into the two bins. But even if this can be accomplished, what criteria should be used for sorting traits into their proper categories? If frequency of occurrence or evolution are not the criteria to use, what else is there? If we use some other criterion of trait "importance," then we might be able to produce a bin of important traits, but what justifies labeling this bin "human nature," and not just a list of traits important in some respects? Such an account will not be surprising and will not show us what the interesting features of our species are, since we are stipulating from the start what makes a trait important or relevant to our nature.

3.2 Kronfeldner's Trait Bin Account

Before we abandon the trait bin approach, let's consider one more attempt at formulating such an account. Philosopher Maria Kronfeldner recently published a book-length treatment of human nature. The concept of human nature that she offers is a trait bin account, but one different from Machery's. For her, "a typical trait is part of human nature if the developmental resources that make a difference for the (abstracted) trait are conserved over evolutionary time by biological rather than cultural inheritance" (2018, 164).

Let's unpack this definition. By embracing typicality, she agrees with Machery that human nature traits must be typical. The same arguments I waged against Machery's majority requirement thus apply. She differs from Machery in focusing on difference-making developmental resources. Some of these resources are conserved over evolutionary time by biological rather than cultural inheritance. If they are, they are human nature by her account. The addition of *difference making* appears to be an improvement over Machery, since it recognizes the fact argued for above: we cannot infer from the premise that something makes an interindividual difference in the expression of a trait to the conclusion that it plays a privileged causal role within the individuals exhibiting the trait.

Whereas Machery sorts traits into those due to evolution versus those due to learning, Kronfeldner is concerned with the conservation of the trait. Is it conserved via biological or cultural inheritance? We saw that the distinction Machery uses does not hold up under scrutiny. What about the one used by Kronfeldner? I maintain that this distinction, too, fails.

The problem with Kronfeldner's distinction between *conserved via biological inheritance* versus *conserved via cultural inheritance* is that these are not mutually exclusively categories. There are many cases in which culture and biology work in conjunction to maintain traits. Take lactase persistence. Babies the world over are born with lactase, the enzyme that allows them to metabolize lactose, a form of sugar in milk. But after weaning, the enzyme is frequently no longer produced. In Europeans lactase persistence is frequent; in Southeast Asia it is rare. Some continents are quite patchy. Western Africa has a relatively high frequency of lactase persistence, East Africa has a lower frequency, and central and southern Africa lower still. Northwestern India has a high frequency, southeastern India a lower frequency (Leonardi et al. 2012).

What lactase persistence is linked to is a history of pastoralism or agropastoralism (arguably in conjunction with disease and famine; see Evershed et al. 2022). In the case of Europeans, indications of milk consumption date back to around eight thousand years ago in northwestern Anatolia and Thrace (Evershed et al. 2008). Lactase persistence has in some cases endured for thousands of years. But how are we to link this with human nature? Is the lactase persistence trait in the contemporary descendants of these early milk drinkers a part of their nature? For Kronfeldner, the answer amounts to asking whether the trait (1) is typical and (2) has been conserved via *biological inheritance* or *cultural inheritance*.

Concerning typicality, the lactase persistence trait is in the minority. It occurs in around a third of the members of our species (Ingram et al. 2009). Thus, it certainly does not meet Machery's majority condition. Kronfeldner's term *typical* is vague, but it seems unlikely that she would consider one third typical. The more interesting question, however, is whether the trait has been conserved via *biological inheritance* or *cultural inheritance*. It is tempting to point to the genetic basis of lactase persistence, which is well known (Swallow 2003). But why did this mutation arise and spread through certain populations? It is due to the milk consumption behavior of the population members (again, in conjunction with disease and famine, factors exacerbating the problems of consuming milk in the absence of lactase), which is maintained via cultural inheritance. So, it may seem that it is instead cultural inheritance that underlies lactase persistence. But, of course, without the genetic mutation arising and spreading, the behavior would not have persisted. Lactose intolerance inhibits the spread of milk consumption behavior.

It should be clear that attempting to point to *either* biological *or* cultural inheritance will cause you to go round and round endlessly. Both forms of inheritance are necessary to maintain the lactose tolerance phenotype. This is not a case of not knowing enough in order to be able to pull the causes apart. Instead, the causes are intertwined and cyclical. Milk consumption is not a unique case. Human language, to take another example, has persisted via an interplay of genetic evolution (on neurological, muscular, etc. traits) and cultural evolution (the transmission and transformation of languages). The failure of Kronfeldner's account to apply to cases of this kind is reason to reject it.

I have argued that neither Machery's nor Kronfeldner's are satisfactory accounts of human nature. But I would like to go beyond this and argue that these are excellent attempts to make the trait bin approach work. Their failures thus reach beyond their own accounts and call into question the trait bin approach in general.

Before setting the trait bin approach aside and developing an alternative, we should pause to consider what possible properties can be used for the segregation of human nature traits. The three key categories of properties are (1) statistical, (2) historical, and (3) intrinsic. Both Machery and Kronfeldner use only (1) and (2). I argued above that statistical properties like trait frequency are subject to a host of problems, chiefly their arbitrariness and their inability to accommodate quantitative traits. While Machery and Kronfeldner both also used historical traits, they chose different ones. As we saw, neither worked, the main reason being that history is complicated and the causes they took to be separable are in fact intertwined. Thus, human nature cannot be based on separating traits into discrete, mutually exclusive categories like *those due only to culture* or *those due only to genes*. I can offer no general proof that no historical property will be a satisfying tool for trait segregation, but it is hard to come up with better alternatives to those of Machery and Kronfeldner – and these failed.

This leaves us with the category of intrinsic properties. These properties are not delimited based on their evolutionary history or their frequency of occurrence. We might, for example, label as a part of human nature traits that arise early in development. But such labeling is always subject to the question of why this property is relevant to human nature. Why would milk teeth be part of our nature, while the later-developing adult teeth would not be a part of our nature?

If we use intrinsic properties for defining human nature, it is hard to avoid simply overlaying our values onto human traits. Traits we think of as important may be segregated into the human nature bin, but this just makes it the bin of traits we find to be important. Human nature then reduces to a mere reflection of our interests.

The conclusion we should draw from the manifold problems with the trait bin approach is that we should not attempt to segregate traits into a human nature bin.

Giving up such an attempt will, of course, bear a cost, and it may seem that the cost is losing the very concept of human nature. For what could human nature be but a special kind of trait? I will now argue that a rejection of segregation is not a wholesale rejection of human nature, and that there is a viable alternative way to understand the concept.

4 The Trait Cluster Account of Human Nature

Gaze up at the cloudless sky on a moonless night far from the polluting glow of cities. You will see countless points of light, some planets, others stars, some whole galaxies or clusters thereof. The lights differ in their color and brightness, and they are not evenly spaced as if set on a grid. Instead, they are clustered, and some salient clusters are given names – constellations such as Aquarius, Canis Major, Cassiopeia, Orion, Ursa Major.

What is the nature of the night sky? If you were a *star bin theorist*, you would say that there is a subset of stars that constitutes sky nature. You might, for example, think that only the brightest stars make up sky nature since they are the stars that can be seen under the widest range of viewing conditions. You might introduce a threshold: perhaps the stars belonging to sky nature are the ones viewable under at least 50 percent of nocturnal viewing conditions. A description of the nature of the night sky would thus simply be a list of the stars that meet or exceed this threshold.

A *star cluster theorist*, on the other hand, will not argue that some stars are part of the sky's nature while others are not. Instead, all stars are a part of the nature of the sky. The star cluster theorist will focus on the clusters, not just named constellations, but clusters at all scales. In this view, a description of the nature of the night sky is a description of the way stars are distributed and clustered across the sky.

What is a better characterization of the sky? Knowing the distribution patterns of the stars across the sky is surely more informative of the night sky than a list of the stars that are frequently seen. What I will argue in this section is that just as we can distinguish star bin and star cluster accounts of sky nature, so we can distinguish trait bin and trait cluster accounts of human nature. Just as the star cluster account is superior to the star bin account, so is the trait cluster account superior to the trait bin account.

The trait cluster account requires a gestalt shift in how we understand human nature. Instead of seeing human nature as a bin of traits, *human nature consists in the relationships among traits*. In particular, human nature lies in the way traits are clustered within and among human life histories.

To motivate this account, let's consider the basic components from which it is built: life histories. Each of us lives a unique life. We experience the world, adapt to it, react to it. And through the course of this life, we exhibit traits. Again, this is "trait" in a very broad sense, including behaviors, psychological states, and morphologies, ephemeral and enduring. While we can all live but one life, there are other possible lives we could have lived. I actually became a philosopher of science, but I could have become a biologist. Had I become a biologist, I would have had many different experiences, which would have shaped my life and would have played a role in shaping how traits are distributed and expressed across my life history.

Now imagine not one alternative path, but all the possibilities. The possible me who lives a happy life into my eighties, the possible me who dies in a car accident as a teenager. If we could map these possibilities, we would see that traits are not dispersed over life histories in a random way; they express a pattern. In some cases, the pattern will be expressed as a stubbornly persistent trait. I am hopeless at spelling, and I don't think any possible me won the US National Spelling Bee. Other traits are predictable, but contingent on environmental inputs. The nature of my facial wrinkles comes from factors such as exposure to UV light, frequency of smoking, and amount of stress endured. Each of these factors is linked to subsequent increases in skin wrinkles. The wrinkles come after and because of these factors.

The patterns exhibited by traits over your set of possible life histories is like the pattern of stars in the night sky. This pattern is unique to you, your individual nature based on all the possible ways your life could have gone. Each of us has such a nature. We can thus define *individual nature* in the following way: *Individual nature is the pattern of trait clusters within the individual's set of possible life histories.*

4.1 The Life History Trait Cluster Account

Let's now consider not just the possibilities of an individual life, not just individual nature, but human nature. Just as individual nature is built from the possible life histories of individuals, human nature is built from the totality of possible human life histories. From this set of life histories, we can define the life history trait cluster account of human nature: *Human nature is the pattern of trait clusters within the totality of extant human possible life histories.*

Let's unpack this a bit. One thing you may find puzzling is that it concerns *possible* life histories for *extant* humans. From the discussion above, it is clear why individual and human nature is based on possible life histories. But why actual and not possible humans? Why allow variation in environmental circumstances but not genes? There are two reasons. The first is that this is a concept of *human nature* in the here and now. This is not about the nature of our ancestors or our possible descendants. While it might be interesting to consider the nature of our ancestors one hundred thousand years ago, or our descendants a hundred thousand years hence, their nature is distinct from our nature. Human nature concerns humans at a particular time, not the whole species from its origins until some unknown future.

The second reason is that if we don't tie "human" to actual humans, then it is difficult to know where to draw the boundary. Which genetic combinations are permissible in constructing possible humans? Do we include any possibility, despite vanishingly low probabilities? If so, and if we are projecting into the future, then the result might be something not recognizably human. Including these strange creatures under the rubric of human nature would reduce, not increase, our understanding of the nature of our species. Just as our species will evolve in the future, it has evolved in the past. Thus, while it was argued above that an individual is a human if and only if it belongs to *Homo sapiens*, the "human" in "human nature" is better considered to be a time slice of our species. If we want patterns of trait expression to be able to characterize – and perhaps even explain and predict – human behavior, then we should restrict human nature to humans that exist now.

Thus, amalgamating all the individual natures from actual humans gives us human nature. Within human nature, there will be patterns of different kinds at different scales. Some will be culturally variable, others will be universal, or nearly so. Not everyone's hair becomes gray, but when it does, it does so gradually, slowly replacing the existing brown, black, blond, or red hair. It is an irreversible process. Short of dyeing your hair, the proportion of gray will only increase. We can think of this graying in terms of changes over life histories. There will be variation within possible life histories (some life experiences can lead to earlier or later graying), and there will be variation across individuals (some genotypes will be more apt to lead to graying). Within each life history, the gray trait will tend to increase, not decrease.

A simple way to think about these patterns is in terms of antecedents and consequents, though more sophisticated analyses are possible. An antecedent trait is one that occurs earlier than the consequent. In this case, *not gray* is the antecedent trait and *gray* is the consequent. The link between the traits need not be perfect. A simple way of quantifying the relationship is to distinguish the *pervasiveness* of the antecedent, which is how commonly it occurs in possible life histories, from the *robustness* of the association, which is how tightly the consequent and antecedent are linked. Think of the robustness of the association as the conditional probability of the consequent given the antecedent.

Some traits will have a low pervasiveness, but have a robust association with a consequent. It is not common to be poisoned by cyanide, but anyone who is quickly succumbs to death. Other associations are more pervasive but less robust. Smoking is fairly common, and lung cancer is associated with smoking, but the association is nowhere near as robust as it is for imbibing cyanide.

Robustness is a metaphysical notion. It describes how traits are actually related, not how we can or should know about this relation. Another way of framing the relation is to consider what information is provided by knowing that a trait is in a particular life history. Depending on the nature of the species and the kind of trait, more or less information about other traits will be provided. Take a pair of traits: if one trait occurs then how much information does it carry about the occurrence of the second trait? Knowing that someone consumed a dollop of cyanide carries a lot of information about what subsequent traits will appear in their life history. Knowing that someone smoked a cigarette at a particular moment carries less information about the subsequent sequence of traits. If knowing that a particular trait occurs in a life history carries no information about prior or subsequent traits, then its occurrence is causally disconnected from the other traits. If all traits were disconnected in this way, we could make no generalizations about humans other than giving a list of traits that they may exhibit. But because the occurrence of one kind of trait carries information about whole cascades of other traits, we are safe in arguing for a trait cluster account of human nature.

The trait cluster account I defend here will be labeled the *life history trait cluster* (LTC) account in order to distinguish it from other trait cluster accounts. Before I further develop and defend the LTC account, let's consider other trait cluster accounts to see how they differ from the LTC account.

4.2 Other Trait Cluster Accounts: Griffiths

Griffiths (2009) offers what can be considered a trait cluster account of human nature. As he puts it: "The primary sense which should be attached to the term 'human nature' is simply what human beings are like, not some cause that makes them that way. As such, human nature is primarily the pattern of similarity and difference amongst human beings" (53). In the first sentence, Griffiths distinguishes cause-based accounts from trait-based accounts. He argues that human nature concerns what we are like, not the hidden causes that bring about our traits. In the second sentence, he implies that human nature is not a bin of traits, but instead concerns relations among traits: patterns of similarity and difference among members of our species.

How is Griffiths's account different from the LTC account? For Griffiths, human nature is about comparisons across individuals, how they differ from one another and how they are similar. These comparisons take place among realized traits of real people. This has the advantage that human nature is readily accessible. It is not hidden in dispositions or unrealized possibilities. While this is a strength of his account, it also has shortcomings relative to the LTC account.

Even if nobody is currently poisoned by cyanide, it is nevertheless our nature to die if we consume it. But if human nature concerns only actual traits, it will in this case exclude perishing by cyanide consumption from human nature. Similarly, even if there are no astronauts in space, it is still human nature to lose bone density while living in zero gravity. Including unrealized traits thus offers a richer understanding of our species.

Another difference from Griffiths's view is the LTC account's emphasis on patterns *within* the life histories of individuals, not just patterns *across* individuals. Griffiths refers to "similarity and difference amongst human beings," but the LTC account adds to this similarities and differences among life histories for individual humans and also the sequence of traits over these life histories. One reason why this is important is that the sequence of trait development over life histories is crucial for characterizing humans. Chimpanzees can't drive cars, at least not well enough to pass a driving test. But let's say we can get one to drive well, but that it takes years of intensive training and drugs and even brain surgery. Would driving a car drop off the list of uniquely human traits? To think so is to focus merely on developmental outcomes, not on the process of development, but it is precisely this developmental trajectory that sets us apart. We easily learn to drive, whereas a chimpanzee, if it could learn to drive, could only do so under a very different developmental trajectory. Our trajectory, the timing of the driving trait and what needs to precede it, sets us apart from all other species. It is the pattern that is important, not (just) the result.

Before moving on, I should point out that in later articles – Griffiths (2011) and Stotz and Griffiths (2018) – Griffiths and Stotz present a view that is closer to the LTC account. They use a developmental systems framework to build an understanding of human nature. One point of distinction is that they are concerned less with trait cluster distributions and more with the causes of these distributions: "While Ramsey focuses on descriptive property clusters that make up human nature, the developmental systems account focuses on the underlying processes that account for these clusters" (Stotz and Griffiths 2018, 68). A detailed comparison of the LTC and their account can be found in their essay.

4.3 Other Trait Cluster Accounts: Cashdan

Let's consider another trait cluster account, that of Cashdan (2013). As is typical of a trait cluster account, Cashdan takes the patterns of trait expression to be of central importance, but she develops her ideas in a way different from that of the LTC account:

> Because human nature evolved to be flexible in predictable ways, the task of understanding human nature requires that we understand how evolution shaped that variation. The assumption is not just that we evolved flexibly, but that selection shaped the nature and direction of that flexibility. To a behavioral ecologist, then, the predictable, patterned nature of that response is the universal we must understand. In this view, we cannot understand our universal human nature without understanding the variability in its expression. . . . The concept is clarified by viewing variation as a norm of reaction – the pattern of expression of a genotype across a range of environments. (71)

As with Griffiths, the focus is on patterns of trait expression, not on identifying a bin of traits. But Cashdan's account differs both from Griffiths and from the LTC account. One important difference lies in her background as a behavioral ecologist and thus her perspective as a biologist, which is seen in her emphasis on the conceptual and mathematical tools we can use to measure and quantify human nature. Her account centers around the *norm of reaction*, a concept introduced (under the label *Reaktionsnorm*) by German zoologist Richard Woltereck in 1909 (Peirson 2012). It is a powerful tool for understanding how phenotypes can vary, sometimes dramatically, given different environmental inputs.

Typically, a norm of reaction takes a genotype and maps how changes in an environmental variable affect the resulting phenotype. Gupta and Lewontin (1982), for example, examined how abdominal bristle number on the fruit fly *Drosophila pseudoobscure* varied with temperature. Individuals in the species typically have between twenty-five and thirty-five bristles. What determines how many an individual has is in part due to their genes, but overall, it is their genes in combination with environmental factors, such as temperature. The interesting finding in this case is that for some genotypes over a range of temperature, temperature and bristle number are correlated, whereas for a different genotype of the same species, bristle number and temperature are anticorrelated.

Thus, it is wrong to say that it is their nature to have a particular bristle number. Nor should we consider as part of *D. pseudoobscure* nature the average change in bristle number over a temperature range. (Consider that if for half the genotypes, there is a correlation between temperature and bristle number, and that for the other half they are anticorrelated, then there could on average be no change in bristle number with temperature.) Instead, what is important is how bristle number varies for each genotype across temperature gradients. Norms of reaction are thus important. They are important to Cashdan's account, but also to the LTC account. However, the role they play in these accounts differs.

Reaction norms generally examine changes in a single phenotypic trait (like bristle number) over a range of a single variable (like temperature). They offer an informative insight into the complexity of gene–environment interactions, and they represent an important feature of how traits are distributed over life histories. But there is more to how traits are clustered over the set of life histories than is captured by norms of reaction. One that was discussed above is the sequence and timing of the appearance of the traits. Fundamentally, human nature concerns how traits are distributed over life histories, but there is no single optimal way of studying or representing these trait distributions. Reaction norms are one, but not the only, way. It is therefore not that Cashdan is wrong about the importance of reaction norms; it is just that reaction norms are but one kind of tool for understanding our nature.

4.4 Trait Cluster Accounts and Processes: Does Everything Flow?

Another account that might be classified as a trait cluster account is that of John Dupré (2018). He states that "if there is a human nature, it does not consist just of a set of properties that humans possess, but of properties that humans possess, or typically possess, at particular stages of their lives" (93). This appears to be a rejection of trait bin accounts and an endorsement of a trait cluster approach. He even notes that "a human is not a thing with a fixed set of properties, but a life cycle" (93). I would reword this – to understand human nature, we must take into account not (merely) static properties (those that persist over whole life histories), but those that vary across life histories – and go on to note that many of the traits are arranged in quite specific ways over these life histories. Further, we should not stop at considering actual life histories, but should consider possible ones as well.

This is not what Dupré does. Instead, he is intent on establishing the "process perspective" for understanding humans. Humans are "better thought of as processes than as things or, in traditional language, substances" (2018, 93). At this point, his argument departs from the path I would take. He asserts that humans are, ontologically, processes. One argument for this is that "the default condition for a substance is stasis. . . . For a process, on the other hand, its very persistence will normally require explanation" (94). I find this to be a strange argument. For one, it is false for many substances and processes. Radioactive substances, for example, have decay as their default, and a lack thereof would require explanation. Perhaps Dupré would argue that such substances are not in fact substances, but are processes, but a reply of this kind appears question-begging. Furthermore, all elements have half-lives, even if some are very long. This implies that over very long timescales, stasis is not the expectation of any element (or any object for that matter).

The bigger problem, however, is the forced choice of ontology: Are humans (or trees or rocks or ingots of uranium) objects or are they processes? The obvious reply to this is to ask what purpose such a classification plays, and why something can't be both. If you sit down at a meal and admire the silverware – picking up a fork and saying to your host, "You have such wonderful objects" – it would be strange to receive this reply: "I polish the silver each week to get rid of the tarnish. That thing in your hand is a process, not an object."

What is true is that a fork is an object for the purpose of my meal, but that it is a process, or a component of a process, for whomever does the weekly polishing. Similarly, a human is an object bearing properties for some purposes, but a process for others. If you want to know how strong to make a car safety belt, you should treat a human as an object with certain properties (flexibility, mass, etc.). But if you are studying aging, it makes the most sense to treat them as a process. No general choice needs to be made.

Where does all this lead Dupré? As he writes, "I also advocate a stronger reason for rejecting the concept of human nature: that I take humans to be better understood as a process than as things or substances" (2018, 104). His conclusion is that "it is safer to dispense with its use altogether" (105). Thus, (1) organisms are processes, (2) processes have no nature, therefore organisms (and the species they compose) have no nature. I think both premises should be resisted. A butterfly is an object with properties (like colorful wing patterns), even though it is part of a complex life cycle, which includes an egg, caterpillar, and chrysalis. Even if we insist that an organism is a process, the life history it realizes is not random, but highly structured. It is populated with traits, many of which have specific sequences that they follow. The nature of the individual is based on these specific sequences.

Thus, while I agree with Dupré that we need to consider whole life histories, I think that this is the basis for a richer understanding of human nature, whereas he thinks that it calls into question the very coherency of the concept of human nature.

Now that we have seen what the LTC account is and how it differs from related accounts, let's consider some possible critiques of this account before examining the roles it can play within and outside of the sciences.

5 Challenges to the LTC Account of Human Nature

The LTC account is not without critics. It has been criticized for being overly permissive and for being too tightly linked to the sciences (and therefore leaving the human out of human nature). While not a direct critique of the LTC account, human nature is often considered to be about what lies within, about our core, not our veneer. Yet the LTC account seems to have no place for the core–veneer distinction. We will consider each of these issues.

5.1 The Permissiveness Challenge

The first challenge we will consider is the permissiveness challenge – the charge that the LTC account, by including all traits, is drained of its usefulness, such as its ability to explain the occurrences of human traits. We saw in Section 3.1 that Machery (2008) offered a trait bin account of human nature. In later papers, he went on to refine and elaborate this account, but also to critique my LTC account. For example, he claims that "because every phenotype that a human being could have belongs to one of the life histories included within human nature, on this notion one cannot justifiably infer that a human being is likely to possess a trait from the fact that this trait belongs to human nature" (Machery 2016, 216). This critique shows one motivation behind his trait bin account. He wants to preserve this inference: *trait X is part of human nature, therefore any human being will probably possess X*. The question to ask here is whether this inference is more or less informative than a statement like *X is associated with traits Y and Z*. Thus, it seems that Machery is interested in statements like *depression is not part of human nature, therefore most human beings are not depressed*, whereas the focus of the LTC account is on trait clusters like *chronic abuse often leads to depression, and depression frequently leads to drug use*. I leave it to the reader to judge which kind of statement is more informative about our nature.

Machery is also concerned with the causal-explanatory function of human nature. He claims that "the causal-explanatory function is largely left unfulfilled by Ramsey's life history trait cluster account of human nature. Every possible trait belongs to some life history included within human nature, so asserting that a given trait is due to human nature provides no information at all" (Machery 2016, 221). It is worth reflecting on Machery speaking of a trait being *due to* human nature. What could this possibly mean within his account? His trait bin account is a way of labeling traits as human nature, and it is unclear what it could mean to say that these traits are also *due to* human nature. If you disassemble your car and label the hundred smallest parts "small," it would not then make sense to then say that a given small part is *caused by* or *due to* its smallness.

By contrast, the LTC account does offer ways to explain actual trait occurrences. To use the above example, if we find that a group of siblings is depressed, we could use the fact that they were abused as children, in conjunction with the human nature fact that depression is associated with abuse, to explain the depression.

Machery is not the only critic of the LTC account. In her book *What's Left of Human Nature?*, Kronfeldner challenges the usefulness of the LTC account. She asserts that the account "treats every trait that a human can possibly develop as part of human nature ... the problem is that it lacks contrastive power. It is too inclusive" (2018, 134). But this is to mistake my trait cluster account for a trait bin account. A trait bin account that included all traits would, of course, be of no use. But it is not that the trait cluster account "includes" all traits, but that it rejects the idea that human nature is about segregating traits into those included in human nature and those lacking this label. The Kronfeldner quote reveals that she has failed to make the gestalt shift from viewing human nature as a bin to seeing it as clusters. Her permissiveness critique thus cuts no ice.

Along similar lines, she argues that "an all-inclusive human nature will result in a situation where the claim that human nature exists is tantamount to the claim that it does not exist. A concept of human nature that includes every possible trait is empirically indistinguishable from the claim that we have no such nature, that we are free in the sense that existentialists like Simone de Beauvoir and Jean-Paul Sartre stressed, a sense that implies limitless freedom" (Kronfeldner 2018, 135). This is a puzzling response. The LTC account equates human nature with patterns of trait associations within and across human life histories. The "claim that human nature exists" would then amount to the claim that traits are not randomly dispersed over life histories, but instead exhibit patterns. Given this, there is no sense in her statement that "the claim that human nature exists is tantamount to the claim that it does not exist." This amounts to asserting that the claim that patterns of trait expression exist is equivalent to the claim that they do not. Think again of the nature of the night sky. It is nonsense to say that because sky nature includes all constellations, asserting that sky nature exists is "tantamount to the claim that it does not exist."

Kronfeldner's claim that the LTC account implies that "we are free in the sense that existentialists like Simone de Beauvoir and Jean-Paul Sartre stressed" is perplexing since the whole point of the LTC account is to understand how we are *not* free, how traits in one part of a life history constrain the occurrence and sequence of traits in other parts of the life history. Furthermore, for the LTC account, our degree of freedom is an empirical question. Given any life history outcome, we can ask how inevitable it is, and what it would take to achieve or avoid it. Thus, there is no sense in which the LTC account "implies limitless freedom."

Critiques of the LTC account like those of Machery and Kronfeldner generally derive from taking it to be a kind of trait bin account, an overly liberal one. The critics fail to see that trait cluster accounts are different in kind. They view the LTC account as different only in how it segregates traits and conclude that it is hopelessly permissive. If we keep in mind that the goal is not to label a bin of traits as human nature, we can see that the critiques miss their target. A trait bin account so permissive that it placed all traits in the human nature bin would indeed be vacuous. Permissiveness is a genuine critique of such an account. But a critique of permissiveness makes no sense for a trait cluster account, since such an account is not in the job of segregating traits.

Human nature understood as patterns of trait expression thus not only fails to "imply limitless freedom," but is useful for understanding ourselves. One way to show this usefulness is to make clear the link between the LTC account and the human sciences. If human nature as understood by the LTC account is at the foundation of the human sciences, then it has clear uses and is not overly permissive.

To show the link between the LTC account and the human sciences – broadly considered, from anthropology and sociology to psychology and neuroscience – we must ask what it is that the human sciences seek, what they hold to be publishable results. Through answering these questions we can see whether the LTC account can serve as the foundation for the sciences. Answering them can additionally help to further assess the strengths and weaknesses of the competing conceptions of human nature we considered earlier. If the human sciences are primarily focused on studying a particular set of traits – the traits identified by Machery or Kronfeldner as human nature, for example – then a trait bin account might be useful after all. If scientists are interested in demarcating traits, in sorting them into those belonging to human nature and those outside of our nature, then there would be reason to consider our nature to consist of a bin of traits. On the other hand, if scientists mainly focus on determining how traits are related to one another, then this is yet another reason to maintain that a trait cluster account is the best way of characterizing human nature.

The challenge, however, is finding the right sort of data to assess whether sciences focus on trait bins or on clusters. One way would be to survey an array of journal articles from the human sciences to attempt to assess this question. I did this previously for the journal *American Psychologist*, the official journal of the American Psychological Society (Ramsey 2018). Psychologists are clearly part of the human sciences, and studying their articles can help to inform us whether they concern how traits are associated or how traits should be delimited into bins.

The result I obtained from that study is that the articles are almost universally about trait associations. They take one trait (meditation, say) and investigate how it is linked to other traits (like stress levels). It would be tedious to go through dozens of articles to attempt to prove my point, but let's examine some articles from a recent volume (74) to get a taste for why it appears that the LTC account fits better with the sciences than a trait bin account.

Issue 1 of this volume is a special issue on racial trauma. The first article (Hartmann et al. 2019) is "American Indian Historical Trauma: Anticolonial Prescriptions for Healing, Resilience, and Survivance." The article considers historical traumas, their causes, effects, and how we should conceptualize them. This is clearly a study of how traits are clustered, how certain oppressive actions can lead to profound effects. The other articles in the issue similarly deal with traumas – like race-based wartime incarceration – and are attempts to understand the impact that such traumas have on the life histories of the traumatized.

Issue 2 begins with the article "The Future of Sex and Gender in Psychology: Five Challenges to the Gender Binary" (Hyde et al. 2019). The article considers the factors that lead to nonbinary gender and "developmental research suggesting that the tendency to view gender/sex as a meaningful, binary category is culturally determined and malleable" (171). This thus considers how traits develop over life histories, not a bin of traits.

Issue 3 is a special issue on multidisciplinary research teams. The first article (Bisbey et al. 2019) argues that "team training contributes to improved performance, reduced errors, and even saving lives" (278), and thus concerns the relationship between a trait (being trained) and subsequent effects.

I could go on – and I encourage you to do so, not just for this journal, but for others in the human sciences. What I expect you to find is that the sciences are not focused on human nature in the sense of Machery or Kronfeldner, or of trait segregation in general. The focus of research in the human sciences overwhelmingly concerns identifying particular traits and finding what traits these are associated with. Such traits might range from genes to cultural experiences: meditation and stress reduction, alcohol consumption and oral cancer, sitting and heart disease. Such associations are studied and quantified, and causal models are offered. Having a concept of human nature that serves as the subject of these studies sets it on an empirical foundation. The LTC account articulates this foundation; it is not overly permissive or vacuous, as has been claimed by trait bin theorists. Instead, it fares much better than their accounts when identifying what it is that the sciences study. Furthermore, the LTC account addresses a key critique of the use of the human nature concept: it can

sometimes promote dehumanization (Kronfeldner 2018). If humans can partici-
pate to a greater or lesser degree in human nature, then one might think that they
vary in the degree to which they are fully human. A trait bin approach can have
this problem, since individuals vary in the number of traits they possess that
belong to the nature bin. But the LTC account has no bins and thus sidesteps this
problem.

5.2 Does the LTC Account Leave the Human Out of Human Nature?

You are hopefully now convinced that the LTC account may be able to serve as
the prime subject of the human sciences. But perhaps you have qualms about the
account, or more generally about linking human nature with the sciences in this
way. Might such an account of human nature be drained of the important
features we would want from it, that the cost of tying it to the sciences is too
great? The anthropologist Tim Ingold, in a review of Elizabeth Hannon and Tim
Lewens's 2018 edited collection, *Why We Disagree About Human Nature*,
argued that "one of the oddities of this book is that while its contributors have
much to say about the concept of nature, and its application to humanity, they
are largely silent about the concept of the human itself" (Ingold 2019). He goes
on to challenge my chapter in particular – a chapter in which I make the case for
linking human nature and the human sciences via a trait cluster account. As he
puts it:

> Now of course, what people do, why and how, can be guided by both intuitive
> judgment and religious prescription. What Ramsey seems to be saying,
> however, is that to study human nature is to set aside whatever people
> might feel about their doings, as well as any moral or religious conviction
> with which they may be freighted, in favour of a scientifically dispassionate
> observation of human behaviour, as it were from the outside. It is to place
> scientists in a realm above and beyond the world they study, immunised from
> any infection that might come from a too close or intimate contact with it.
> Now of course it is precisely this closeness and intimacy – this feeling for the
> world of which we are intrinsically a part, and to which we owe our very
> existence as living beings within it – that lies at the heart of intuitive or
> religious sensibility. Ramsey's declaration, in writing off such sensibility
> from any study of what it means to be human, amounts in effect to
> a defence of normal science, and of the absolute separation of knowing
> from being on which it depends. Yet are not practitioners of the human
> sciences – among whom Ramsey includes psychologists, sociologists,
> anthropologists and economists – also human beings themselves? And as
> such, are they not inevitably embroiled in responsibilities towards those
> among whom they work and study, and whose ways they seek to understand?
> (Ingold 2019)

There is much to unpack here. To start with, Ingold seems confused about the trait cluster account. The account does not imply that human nature is independent of intuitions and of religion. In fact, our intuitions, imaginations, and creativity are central features of our nature. Recall the liberal sense of "trait" employed by the LTC account. Acts of creativity, devotion, and imagination are included within the set of traits.

What I am arguing is that imagining human nature to be a certain way does not make it so. If we imagine that humans have an evil core placed in our souls by the devil, that does not make it so. We have intuitions about the size of the Sun and its relation to the Earth. The Sun appears to be smaller than the Earth and to travel across the Earth each day like a glowing rock slowly traversing a viscous fluid. Is this a veridical representation of the nature of our solar system? It is not, and science offers a corrective.

Most religions are founded on a creation myth about how the Earth came about and how landforms were created. The Bible tells us of a massive flood that wiped out much of life on Earth and shaped the landforms, placing marine fossils on hilltops. Should we derive the nature of the solar system (or universe for that matter) from the Christian Bible, the Mahabharata, or other religious texts? It is science that offers correctives; it allows us to see that the world is billions of years old, not thousands. Just as our solar system is a structure of lumps of matter moving in predictable ways, human nature is a structured set of traits distributed in predictable ways over life histories, appearing and disappearing and transforming over the span of these histories. It is this to which we can turn our scientific eye in understanding human nature. It is this that human scientists are studying when they study humans.

Thus, pace Ingold, moral and religious convictions are not left out of the LTC account. To have a conviction is to possess a trait, and we can inquire about which traits are associated with convictions of various kinds. Convictions, intuitions, and faith are not excluded from human nature, but they are not privileged either. They are treated as traits among countless others. What is excluded is the thesis that human nature is fundamentally based on faith or intuition, that it is immune to scientific scrutiny.

The human sciences clearly study humans. The question for Ingold and others skeptical of the LTC account is whether they study *human nature*. If we answer yes, then human nature can be understood as a (or *the*) subject of the human sciences. If we answer no, on the other hand, then human nature is something outside of the sciences as currently practiced. It may seem that if human nature is unmoored from the sciences, this is liberating. But such liberation comes at a steep cost.

Suppose we want human nature to concern what humans ideally are, not how we actually are. As we saw in Section 2.1, such a human nature concept would not be discovered through observing how we are built and how we behave, but by asking humans what it is to be an ideal human. Such a conception would cause unresolvable divisions and disagreement. There is no consensus on how we should be. Basing human nature on our intuitions about how we should be, or on religious texts that dictate how we should be, would be a mere repackaging of these intuitions or decrees, an attempt to give them more authority. *It is not merely good to be generous to strangers, it is our nature to do so*, one might claim. But if human nature is unmoored from the sciences, such a claim would amount merely to asserting that we find it to be especially important to do so. It would not be a claim about how humans are built.

Sometimes human nature is viewed in the opposite way, not as an ideal to strive for, but as a challenge to overcome. We might think that we are by nature selfish, but that we should aim for altruism. That we are by nature promiscuous, but should strive for fidelity. Human nature in this sense is the collection of raw instincts that we need to overcome through cultural mores and strength of will. There will be more to say about problems with the idea that we have a natural core coated in a cultural veneer in Section 5.3. But the key observation at this point is that such an account of human nature would be an attempt to repackage our desires in order to give them the air of objectivity. Because there will be countless and often conflicting ideals, human nature understood in this way would not be a way of increasing the understanding of our species, but of engendering conflicts over incompatible visions of whom we should be, or of controlling and dominating others. I therefore hold that human nature is best taken to be tied to the human sciences. I am not arguing that the human sciences should study human nature, but that they already do so. The observations in Section 5.1 support this.

Next, let's consider Ingold's assertion that I am defending "normal science" through my account. If by normal science he means the concept developed by Kuhn (1962), and held in contrast with revolutionary periods in science, then it is clear that I am not defending normal science since I am in no way in dialogue with the Kuhnian framework. Thus, his invocation of "normal science" is puzzling and seems disconnected from the arguments for the LTC account.

Finally, regarding Ingold's point that those who study human nature are themselves human beings and thus "inevitably embroiled in responsibilities towards those among whom they work and study, and whose ways they seek to understand," it seems that this is just the truism that studying humans is complicated. Studies can be beset with conceptual and ethical issues, and there are deep challenges for any species attempting to study itself. These are general problems, not ones specific to the LTC account, but to all reflections on ourselves.

5.3 Isn't Human Nature about Our Core Instead of Veneer?

Another concern one might have about the LTC account is that it doesn't appear to support the idea that human nature concerns our core, what we are really like, and not our veneer. It seems impossible to carve off the veneer from the core when the material we have to work with is a complex skein of life histories. But how then can we answer questions like: Are we good or bad, selfish or altruistic, at our core? Is it our nature to steal and cheat, but we are held in check by the laws, rules, and norms governing civilization? Or is civilization a corrupting influence, causing us to go against our nature?

The Latin proverb *Homo homini lupus* says that humans are a wolf to others of their kind. This, of course, is taking an unfair view of wolves, which are highly social and cooperative, not at all an example of all against all. Setting this caveat aside, the proverb represents a common way of seeing humans as being aggressive and brutal at their core, with civilization acting as a governor and suppressor. Take, for example, this excerpt from Sigmund Freud's *Civilization and Its Discontents*:

> Men are not gentle creatures who want to be loved, and who at the most can defend themselves if they are attacked; they are, on the contrary, creatures among whose instinctual endowments is to be reckoned a powerful share of aggressiveness. As a result, their neighbor is for them not only a potential helper or sexual object, but also someone who tempts them to satisfy their aggressiveness on him, to exploit his capacity for work without compensation, to use him sexually without his consent, to seize his possessions, to humiliate him, to cause him pain, to torture and to kill him. *Homo homini lupus*. Who, in the face of all his experience of life and of history, will have the courage to dispute this assertion? (1962, 58)

Can we have the courage to dispute this assertion? We need first to understand what such an assertion assumes of human nature. It appears to view humans as vicious, murderous creatures whose core violence is kept in check with a veneer painted on by the pacifying strictures of civilization. What should we make of this core–veneer distinction?

The idea Freud described as the beast within has been characterized by primatologist Frans de Waal (2005) as *veneer theory*, the supposition that we have a selfish, venal, corrupt, even murderous core, and that over that core culture has painted a thin moral veneer. All our cooperative, friendly, altruistic behavior is due to this fragile veneer, a veneer that we can easily scratch, exposing the beast within: "Scratch an 'altruist,' and watch a 'hypocrite' bleed" (Ghiselin 1974, 247).

De Waal contrasts this with the view he endorses, under which we are cooperative and moral at our core. Sensitivity to fairness, for example, does not have its source in culture, but is something at our core. One line of evidence for this is anatomical: the parts of the brain involved in moral emotions includes ancient regions, regions that distantly precede culture and civilization. Veneer theory, by contrast, seems to predict that morality would be centered in the neocortex, the most recent addition to our brain. (The neocortex does, however, play important roles in behavioral inhibition, in getting us to reflect on and suppress problematic urges.) Another line of evidence is comparative: if we examine our primate relatives, some of them seem to exhibit behaviors that, if not full-blown moral, are at least proto-moral. They certainly seem to be sensitive to issues like fairness.

One of de Waal's favorite examples of fairness in nonhuman primates comes from a series of experiments with capuchin monkeys at the Yerkes Primate Center. In one experiment, researchers played a game with the monkeys (Brosnan and de Waal 2003). The game involved handing a monkey a small rock, and if the monkey returned the rock, it received a reward. The reward was either a cucumber slice or a grape. The monkeys like both, but they prefer grapes to cucumbers. If they are alone, they will happily play the game with cucumbers. But things get interesting if there are monkeys in neighboring cages and you play the game with grapes with the first monkey, then attempt to play the game with cucumbers with its neighbor. If this happens, the second monkey will not only be upset about receiving cucumbers, but will often refuse to eat the cucumbers, at times throwing them back at the researcher.

Capuchins, in one interpretation of this experiment, are sensitive to inequity and respond negatively to situations in which they are not treated fairly. Whether this is indeed inequity aversion or just being upset about getting cucumbers when there are grapes around is another question – some researchers have argued that capuchins do not exhibit inequity aversion (McAuliffe et al. 2015), that "although the sense of unfairness, or inequity aversion, seems an immediate and natural reaction, we know very little about its underlying psychological mechanisms" (Dubreuil et al. 2006, 1223).

Whether or not the capuchin studies are correctly interpreted as inequity aversion, it appears that such aversion – and a general sensitivity to fairness – develops spontaneously in young children. Some studies suggest that sensitivity to fairness begins to arise soon after their first year of life (Geraci and Surian 2011; Sloane et al. 2012). In one study demonstrating this effect, Sommerville and colleagues (2013) showed twelve-month-old and fifteen-month-old children videos of two adults given crackers by a third adult. In the equal outcome scenario, each received two crackers, and in the unequal outcome scenario one received one cracker and the other three. The researchers recorded how long the infants looked at the outcomes (which were presented as still images). Twelve-month-olds looked at the even and uneven distribution scenarios equally, but fifteen-month-olds looked considerably longer at the uneven distribution.

Other studies show a similar development of attention to fairness, including equal distributions among equal subjects, but also merit-based distributions – more for subjects who have earned it through, for example, completing a task (Blake et al. 2014). The further development and refinement of fairness concepts and behaviors vary, however, from one culture to another (Rochat et al. 2009; Blake et al. 2015).

Do such studies show that we are at our core averse to inequity, that when we argue for fair treatment, such urging stems from our nature? In the *Scientific American* blog, psychologists Katherine McAuliffe, Peter Blake, and Felix Warneken seem to imply that this is true:

> There is a commonly held belief that humans are fundamentally selfish agents and fairness is a construct designed to help us override our selfish instincts. Not only this, but the idea really seems to be that fairness doesn't come naturally, which is why we need institutions like the justice system to make sure that fairness prevails. Psychologists and economists have begun to gradually chip away at this notion, showing that people are actually pretty fair even when they can get away with selfishness. (2017)

It is false, they argue, that fairness does not come naturally. Instead, it naturally arises, or as trait bin theorists may frame it, it is a part of our nature.

Debates over which dispositions come naturally or are at our core – and which are a mere veneer – presuppose the coherency of the core–veneer distinction. Let's step back to consider the concepts at play in these debates to see if they are coherent. We have two key distinctions. One is to have something at one's *core*, as opposed to *veneer*. Another is to do something *naturally* or *spontaneously* or *instinctually* instead of having it *imposed* or *learned*. And, of course, there is the question of how these distinctions link up with the concept of *human nature*. The temptation is to think that human nature concerns what is at our core, what we do naturally or spontaneously.

In assessing whether this is a line of thought worth pursuing, we need to analyze these distinctions. Recall one of the problems with Machery's (2008) account. He argued that human nature is due to evolution, as opposed to being learned. But what we saw was that the evolution–learning distinction is problematic. (See Section 3.1 for the arguments.) Similar arguments can be waged against the natural/spontaneous versus imposed/learned distinction. Zooming out to view entire life histories, traits can arise at a range of places during a life, and they can have a whole spectrum of possible antecedent inputs. That is, there is a continuous variation in the degree to which the environment, or "experience," plays a role in determining the form and timing of trait expression. There is no nonarbitrary point at which a trait shifts from being spontaneous to being learned.

Furthermore, recall from Section 3.1 that while we can point to particular genes in explaining *differences* between individuals in their trait expression (the difference between having blue or brown eyes, say), this does not mean that we can infer that in an individual the gene that underlies the difference plays a privileged causal role in the production of the trait. Any particular gene is but one element of a causal nexus. The gene is not "for" a trait just as gasoline is not "for" a car starting. (It is a necessary but not sufficient factor in the car starting, and the gasoline is primarily for doing work once the engine is running.)

Similar problems arise for the veneer–core distinction. Culture is not a cloak we wear that we can doff in order to see the nature beneath. Nor are nonhuman primates naked humans. Humans without culture is like water without oxygen. What remains is not oxygen-free water; it is hydrogen.

You may wonder, surely we can gain information about the core not through peeling away culture but by observing when and how behaviors come about. If they come about early, perhaps they are more deeply lodged in our core. The problem is that there are many traits that come about later in life – for example, getting wisdom teeth or gray hair – that seem far from cultural whims. Also, the heritability of some traits increases with age. Bouchard (2013), for example, describes the Wilson effect: the increase in the heritability of IQ with age. It appears that IQ heritability increases until the late teen years before leveling off.

If being early in a life history does not necessarily indicate that a trait is at one's core, perhaps being widespread across our species indicates core as opposed to veneer. For some cases, this appears to make sense. It may seem that it is not in our core to speak French, but it is to speak a human language. This view, however, appears to have problems. If learning French became so common that almost everyone in the world could speak the language, would it then be a part of our core? Is playing soccer part of our core and not veneer because it is played by so many the world over? If core–veneer is a distinction about individual development, then changes in the distribution of traits – like the spread of French or soccer – should not change the classification of the trait from core to veneer or vice versa.

Another possibility for distinguishing veneer from core is to point to the cultural or environmental inputs that go into producing a behavior. If the behavior is due to cultural inputs, it is part of the veneer and not the core, or so one would think. The problem here is, again, that the evidence for this comes from differences among individuals, not individual life histories. The same problem of inferring individual causal processes from interindividual differences applies.

I hold that the idea of a beast at our core overlain with a moral veneer is problematic, not because of what it places at our core, but because of the assumption that there is a clear core–veneer distinction. Like the innate–acquired distinction, the core–veneer distinction likely causes more confusion than clarity (Griffiths et al. 2009). We can observe when traits appear along life histories, what inputs are necessary for their appearance, and the degree to which they vary by culture, gender, race, and such. Because the appearance and variation of the traits come in degrees, they do not point to an isolated veneer that we can carve off from some unmalleable underlying core.

If the core–veneer distinction is problematic, does that mean that we must discard the intuition that our nature is what is within and not what is imposed? Could we perhaps hold on to the idea of a nature within?

5.4 Isn't Human Nature about What's Within?

One of the desires behind talk of human nature is to gain insight into what is at the basis of our species, to peer into our soul, so to speak, and see what we are really like. The LTC account argued for here seems to include environmental inputs, and therefore to be not just to about what is within us. Shouldn't human nature be restricted to what's within?

Robert Greene, in the introduction to his *Laws of Human Nature*, defined human nature as "the collection of these forces that push and pull at us from deep within" (2018, 3). When he described his book, he noted that "each chapter ends with a section on how to transform this basic force into something more positive and productive, so that we are no longer passive slaves to human nature but actively transforming it" (6). In his conception, human nature is like an inner agent grasping at the levers of control, but through an understanding of this agent, we can come to tame it, to have some control over it.

To analyze Green's conception of human nature, it is helpful to consider the idea of a unified interior that causes and explains our behavior. This generally goes by the name of *character*. The idea behind character is that it allows us to make general statements about individuals and to gain insight into them. One way of interpreting Greene's understanding of human nature, then, is that human nature is that part of our character that is universal to members of our species. It is what cuts across individuals and the variations they exhibit in their character.

If human nature is part of our character, we can begin to assess Greene's conception of human nature by first assessing the idea of character. Do we actually have characters? If so, this would give some hope of building human nature from it. If not, then this idea of human nature can be set aside.

To begin, we should ask what it is that would count as evidence for character. One source of evidence would be that we can accurately predict the behavior of an individual in one context based on their behaviors in other contexts. This is what psychologists label *cross-situational consistency in personality* (Shoda et al. 2002). A low level of cross-situational consistency is evidence against character, whereas a high level is evidence for character, the idea being that high cross-situational consistency is best explained by an underlying character. With high cross-situational consistency, we could label someone as *messy*, and this label should apply broadly. If, instead, people are messy in one domain (their bedroom) but not others (their office or kitchen) then the "messy" label is not very useful in explaining or predicting behavior across domains. If there is character, it should play the role of controlling a wide range of behaviors, leaving its mark in the cross-situational consistency of the behavior (Shoda et al. 2002).

The literature on character is large and complex, but one generalization that can be made is that strong cross-situational consistency in personality does not seem well supported by the evidence (Mischel et al. 2002). Instead, behavioral responses are difficult to generalize. Your friend Mika may be polite and deferential at work, but a maniac behind the wheel of her car. She may be frequently late to work, but punctual at social gatherings, or vice versa. Being habitually late or punctual within one domain does not allow accurate predictions of punctuality in other contexts.

What is true, however, is that while there is often little consistency *across* domains, there is frequently consistency *within* them. Shoda and colleagues (2002) describe this in terms of "stable IF-THEN signatures" (324). In this case, we can make statements like IF Mika is in her office THEN she is polite, and IF she is in her car, THEN she is a maniac. The stability, they argue, exists but is local. There is no overarching character, no wizard behind the curtain making sure our behavior is consistent across domains. Instead, consistency arises only when we specify the details of the circumstances.

There is debate over the degree to which one's character or situation can predict behavior (see Doris's 2002 book and the controversy it stirred to get a sense of this discussion). Nevertheless, the extreme view of a powerful character that drives cross-situational consistency is not well supported by the research. In the absence of such cross-situational consistency, we should not posit a unifying character hidden deep within. Instead, we are fractured selves; all we have are stable IF-THEN signatures or, in the language of the LTC account, stable patterns over sets of life histories. And just as this is true of individual character/nature, so it is for human nature.

Does this mean that we lack the "forces that push and pull at us from deep within" (Greene 2018, 3)? We do have forces, but they are not always generalizable in the way that Greene assumes. Life histories do not unfold willy-nilly, but are ordered. Certain traits will trigger the appearance of others. In some domains, generalizations like those of Greene are possible; in others they are impossible, and all we have are stable IF-THEN signatures. What the LTC account argues is that there are stable IF-THEN signatures not just in the domain of social psychology, but across human traits in general. In identifying them, we should not look only for hidden forces, but include all traits along human life histories, whether hidden within or exposed.

6 What Can We Do with the LTC Account of Human Nature?

With the defenses of the LTC account behind us, we should now ask what we can actually do with this account of human nature. What uses might it have? Kim Sterelny (2018) suggests that "we can reject intrinsic essentialism about species in general, and our species in particular, and retain a concept of human nature. But it is a descriptive, 'field guide' concept: it does no explanatory work and is a somewhat arbitrary, list-like conception. Do we need it?" (123). The descriptive field guide account is a trait bin approach, a mere list of human traits that work to generally characterize us. We walk on two feet, wear clothing, have rather sparse fur, speak to one another, and so forth. Such characterizations might help an alien species pick humans out of a lineup of vertebrates, but they don't seem very insightful for those already familiar with our species, and don't seem to explain our traits.

If the choice is indeed between a list of general characteristics and of doing away with the human nature concept altogether, then it seems that we should do away with it. We already know how to pick humans out from a crowd – no field guide is needed. But the LTC account is not a field guide. To what degree can it be more useful than the account dismissed by Sterelny? What can such a concept of human nature be used for? Can it be used to explain the occurrence of behaviors? Can it tell us how we differ from other creatures, from chimpanzees or Neanderthals? Can it bridge uses of "human nature" in the media and the sciences? Are there moral implications we can draw from human nature as characterized by this account? These are some of the questions this section addresses.

6.1 Can Human Nature Explain Human Traits on the LTC Account?

Can natures explain particular trait occurrences? We saw in Section 5.1 that Kronfeldner and Machery challenged the explanatory ability of the LTC account, suggesting that it has taken the easy way out, packing all our traits into human nature. Everything that anyone does is human nature, therefore human nature can't explain behavior. But is this really true?

Let's take a similar argument about the nature of physical objects. Nothing physical objects do is against their nature. Some stars turn into white dwarfs, some into neutron stars, others into black holes. It is not that black holes or white dwarfs, or any other possible outcome of a star's evolution, are a part of the star nature bin. Nor are they outside of star nature. Recall that the inside–outside distinction for the nature of things is problematic and not useful. Instead, consider the life history of a star. The life history trajectory involves a sequence of traits, the nature of which is determined by factors such as the mass of the star. A massive star will turn into a red supergiant, which will supernova, and from there, depending on its mass, it will form either a black hole or, if not sufficiently massive, a neutron star (Kippenhahn et al. 1990). If it becomes a neutron star, it will in old age become a pulsar. It is the nature of stars to follow these trajectories. Neutron stars follow instead of precede supernovae; black dwarfs follow instead of precede white dwarfs.

We can now see how the LTC account can explain the occurrence of one trait based on the occurrence of others in conjunction with the nature of the system in question. In the case of star evolution, we can explain why one trait occurred instead of another, why the star evolved into a black hole instead of a neutron star, based on other properties, in this case the most important one being its mass. It is the nature of stars above a certain mass threshold to form black holes instead of neutron stars, and the occurrence of a black hole or neutron star can be explained by this nature.

The same is true of human traits. Socrates's death is explained by his consumption of hemlock. It is our nature to die from its consumption. We could go deeper than this, pointing out that hemlock contains coniine, and that coniine consumption leads to muscle paralysis, that among the muscles affected are those involved in respiration, and that with all but the shortest absence of respiration, humans die.

If we had a different nature, if coniine had no impact on our muscles, then it would not influence our breathing, not unless it had other effects. If we were able to go for long periods of stasis without breathing, we might be able to avoid death, or at least increase the lethal dose of coniine. But given our nature, a dose of coniine smaller than a single chocolate chip will kill us. Many other plants are harmless to us, but toxic to other creatures. Eaten in moderation, raisins cause no harm to humans, but they can kill a dog. The difference between our nature and the nature of dogs explains this.

Trait cluster accounts, and in particular the LTC account, are thus able to explain particular trait occurrences based on our nature. This is not the case with trait bin accounts, or at least not in a nontrivial way. If we specify, as did Machery (2008) and Kronfeldner (2018), that human nature traits must be traits that exist in the majority of humans, then we might offer this sort of explanation: if someone has the trait, we could attempt to use human nature to explain the trait's occurrence, since if it is a human nature trait, it will be common, and if it is common, it will probably be present in most individual humans taken at random. But it is easy to see that such a putative explanation is trivial.

Similarly, Machery's "result of evolution" criterion offers little ability to explain the occurrence of traits. Although we saw earlier in this Element that this criterion has deep conceptual problems, if there were such a criterion, then if you identified a trait as a part of human nature, you could explain other features of the trait based on this. For such an explanation to work, there would have to be generalizations that can be made about evolutionary traits. You might think, for example, that if a trait is evolutionary, that it is therefore likely to arise in individuals spontaneously, or that it has a genetic basis. To the extent that these generalizations about evolutionary traits can be made, we can explain the occurrence of those associated traits based on the trait belonging to human nature. In sum, if the trait is identified as human nature, this implies that it is evolutionary, which means it will bear properties common to evolutionary traits (whatever those may be).

Trait bin accounts, therefore, are either completely devoid of explanatory power or the explanations they furnish are of dubious interest. They are based on stipulations, the stipulated criteria used to place traits into one bin instead of another. And these stipulated criteria are likely to be of little interest to science. The explanations from the LTC account, by contrast, are precisely those of interest to the human sciences.

6.2 Does the Study of Twins, Triplets, and Other Multiples Give Us Insight into Our Nature?

One potentially useful domain for a concept of human nature is studies of twins, triplets, and other multiples. Can such studies shine a light on human nature? If so, how?

At subalpine and subarctic tree lines, there are often short, gnarled trees called krummholz. We might ask of such trees whether they have the same natures as the larger trees at more southerly latitudes or lower altitudes. What a botanist will do to answer this question is to take seeds from krummholz and from others of the same species and grow them in a common field. If the seeds from krummholz grow into krummholz despite their more hospitable environment, their nature is indeed different.

This example highlights the fact that we can consider the nature of individuals and of the species to which they belong, but we can also consider the nature of any set of individuals. Just as we can ask about the nature of gnarled subalpine fir trees, so we can inquire into the nature of any human group. We can consider the nature of your immediate family, or of the individuals from your hometown, or of Amish women in Indiana, or teenagers from Chicago's south side.

With humans, we cannot normally conduct common garden experiments like that of the krummholz. But there have been natural or deliberate human experiments testing questions such as *to what extent and in what ways do differences or similarities in genes affect individual nature?* One striking example is a study facilitated through the Louise Wise Services adoption agency. In the late 1950s, the agency formed a policy of separating twins and triplets and placing them in separate families. They did not inform the families of the existence of the adoptee's genetically identical siblings (Hoffman and Oppenheim 2019). As part of the study, the babies were intentionally placed in families that varied culturally and in socioeconomic status. The study was designed to offer insight into human nature. Did it do so within the framework of the LTC account?

Take the case of the male monozygotic quadruplets born to a teenage mother on July 12, 1961. One of the quadruplets died at birth, but the three others survived and were placed under adoption by the Louise Wise agency. Each of the boys was placed into a family that had adopted a daughter from the agency two years prior. The families were not told of each boy's siblings. Under the instructions of psychiatrists Viola W. Bernard and Peter B. Neubauer, one boy (who became David Kellman) was placed in a blue-collar family, another (who became Eddy Galland) was placed in a middle-class family, and the third (who became Bobby Shafran) was placed in an affluent family.

Before considering what such an experiment can tell us of human nature, we should pause for a moment to reflect on the nature and ethics of such a study. At the time, it was considered better for both the children and the placement families for multiples to be broken up, and adoptions were generally closed. Thus, the practices of the agency were not radically different from others of the time (Hoffman and Oppenheim 2019). Nevertheless, it is now recognized that such separations are harmful, and that the publication of the results of the study could cause further harm. It is for this reason that the data are sealed in an archive at Yale University until 2065 (McCormack 2018).

Let's now consider what such an experiment could tell us about human nature. As observed previously, each human has a large set of possible life histories, but only one is realized. The unique feature of monozygotic multiples is that they let us get close to seeing more than one possibility realized. The starting conditions are so similar that each multiple represents a possibility of the others. Because of this, monozygotic multiples provide information about *individual nature*.

Thus, we know more about David Kellman's nature through observing the life of Eddy Galland and Bobby Shafran. We could infer that had David been raised in an affluent family, he would have ended up more like Bobby. Of course, such inferences require it to be the case that it is *affluence in general* that explains some of the differences between David and Bobby, not just the particulars of Bobby's parents (who happen to be affluent, though their affluence may or may not be a dominant part of who they are). This complication aside, it is safe to say that the studies do provide information about the individual nature of the study subjects.

But what, if anything, do such studies tell us about human nature? Because human nature is a composite of all individual natures, we automatically know something about human nature from learning about individual nature. What is missing from such a study is knowledge of the extent to which we can generalize. The problem is gene–environment interaction. As we saw earlier in this Element when discussing Cashdan's account (Section 4.3), if we plot norms of reactions for traits, we can see that different genotypes react differently to their environmental inputs. One *Drosophila* genotype may have bristle number correlate positively with temperature over a particular range, while a different genotype of the same species may have a negative correlation over that same range. Thus, for Bobby, David, and Eddy, it could be that affluence has a particular effect over some affluence range for their specific genotype, but it does not follow that this is true of other genotypes. (And of course, affluence is not a simple trait like temperature and is not unproblematically reduced to single variables like family income.) Thus, while such morally dubious studies are compelling on the face of it, without knowledge of the gene–environment interactions at play, they do not tell us much about human nature.

6.3 Can the LTC Account of Human Nature Tell Us How We Differ from Other Species?

In 1931, the seven-month-old chimpanzee Gua was acquired by Luella and Winthrop Kellogg to be raised alongside their ten-month-old son Donald (Kellogg and Kellogg 1933). This was the first of several attempts to raise chimpanzees as humans to see how similar their behavior would be to their human counterparts. Why do such studies? One reason is to test behaviorism, the psychological view that takes behavior as being highly flexible and readily changeable via conditioning. Thus, so the idea goes, if we raised a chimpanzee with the same conditioning as a human, they would behave similarly to humans.

The result of such studies showed that while chimpanzees can have their behavior significantly modified from their wild counterparts, it never approaches that of humans.

But what do such studies tell us about chimpanzee or human nature? The studies deliberately attempt to raise chimpanzees as if they were human. Why not just observe chimpanzees in the wild and note their behavior? Why attempt to give them experiences like that of their human counterparts? One possible answer is that by observing chimpanzees with humanlike experiences, we can get a true picture of their nature. But this is misguided. It is not that chimpanzees raised by humans give a truer picture of their nature, nor is it that their behavior in the wild is a truer picture of their nature. Both are insights into chimpanzee nature, since they reveal possible chimpanzee life histories.

Recall that a species's nature concerns the relationships that hold between traits at one part of the life history of the individuals composing the species and the antecedent and consequent traits with which they are associated. It is plain that chimpanzees in the wild behave quite differently from suburban humans. But to what extent is this difference merely due to the different environmental inputs into chimpanzee versus human life histories? An ideal test of this would be to let chimps raise humans and humans raise chimps. If chimp-raised humans behaved just like chimps and human-raised chimps behaved just like humans, then the behavioral nature of chimps and humans would be identical – within, at minimum, the domain tested. The degree to which the species differ is the degree to which their natures differ. Why these tests are important is that we are exposing the two species to the same antecedent conditions and observing the extent to which their consequent traits differ. This allows us to see how patterns of trait expression differ across the species.

Do chimpanzees and humans differ in their natures? Absolutely. A chimp-raised human would not live long, and a human-raised chimp behaves little like its human counterparts. The LTC account provides a framework for saying precisely how their natures differ. It is not based on a bin of traits exhibited by wild chimpanzees and modern humans, but is instead based on how each species develops over their life histories in response to environmental inputs and prior life history states.

In addition to human-chimpanzee comparisons, we may also want to compare our nature with that of our ancestors or other ancestral sister species. For example, we may wonder whether – or better, to what extent – Neanderthals share our nature. How can we address this question?

It may seem that the LTC account simply says that this question is unanswerable: a species's nature is time indexed, and thus you cannot form comparisons across time. But this response fails to understand what we are asking when we ask how Neanderthal nature compares to our own. If we are asking about the differences between Neanderthal nature and our nature, we are asking counterfactual questions: What would humans be like were they to be raised by Neanderthals? Perhaps more interestingly, what would Neanderthals be like were they to be raised among modern humans? Would they come to speak a human language; would their youth enjoy ice cream, rock music, and playing video games? Could they develop into accountants and grocery cashiers and hair stylists and pilots? What would their limitations be and how would they excel? The degree to which Neanderthals would differ from our species in this domain is the degree to which their nature differs from ours in this domain.

If dendrologists wish to use tree rings to compare the nature of different species, they seek trees from the same region, and when they do their comparisons, they align the rings based on known dates. This way differences in the rings will tell us more about the differences in tree nature. If a particularly hot summer is associated with a narrow ring in one species but a fat ring in another, it may be that the former could not handle the heat and that its leaves wilted in the sun, while the latter might be more tolerant of the heat and could use it to its advantage for rapid photosynthesis and growth. The dendrologists are thus making comparisons among consequent traits (like tree growth) in situations in which the antecedent traits (annual rainfall, temperature, and so on) are similar. The same is true for comparing human nature with that of other species. We anchor the life history comparisons to similar antecedent experiences and then note how the consequent traits develop, providing insight into the nature of the species.

6.4 Can the LTC Account Make Sense of Human Nature References in the Media?

In Section 2.3, we saw that the concept of human nature is used throughout the media, making its way into the titles of popular books and newspaper and magazine articles. Let's now return to those *New York Times* articles mentioned there to see the degree to which their use of human nature can be made sense of in light of the LTC account. John R. Quain's (2016) "Makers of Self-Driving Cars Ask What to Do with Human Nature" concerns the problem of Level 3 automation technology, which is self-driving technology that occasionally requires humans to take control of the car (as opposed to Level 4, which is fully self-driving). The central question is: "Is it possible to get a driver to safely take back control of a car once the vehicle has started driving itself?" It turns out that humans are surprisingly slow in taking control in Level 3 vehicles, exhibiting "an average of 17 seconds to respond to takeover requests. In that period, a vehicle going 65 m.p.h. would have traveled 1,621 feet – more than five football fields." Such studies focus on an input (the car making a noise and/or a visual cue) and studying the subsequent human reaction. What is important is finding out what kind of input is (1) not terribly annoying or startling, yet (2) prompts the user to take control, and (3) does so quickly. Such an investigation is clearly a study of how traits are distributed over life histories, and thus, invoking human nature in such a context fits well with the LTC account.

In Lipkis's (2017) letter to the editor, he argues,

> What causes more murders in today's chaotic world? Is it guns, bombs or cars? Actually, I'd say it's none of the above; it is human nature. Broken families, poverty and poor mental health make people more susceptible to accepting false beliefs and evil ideology. Murder can ensue whether you have a gun, knife, hammer, bomb or vehicle. . . . People will find a way to kill others with or without guns.

This argument is about human life history trait clusters. What is indisputable is the relatively high murder rate; what is less clear is what antecedent traits tend to lead to this. Lipkis is arguing that the easy accessibility of guns is not the most important determinant of these killings. It is instead certain traits associated with the upbringing of the murderers – like poverty, familial problems, the lack of a sound mind and solid education – that are most responsible for the murderous behavior. He goes on to point to the fact that while Americans have by far the most guns per capita, they are far from having the most murders per capita. (It should be pointed out that looking merely at quantity of guns per capita across countries will not give a clear picture of the causal antecedents to gun murders. It would be better, for instance, to examine murder rates in neighboring states that have large cultural overlaps but differ in laws controlling gun availability.)

Whether or not Lipkis's argument is well supported by the data, it is clear that he is using human nature in the trait cluster sense. By blaming human nature, he is not blaming a particular trait. Instead, he is identifying developmental factors that, if they are realized as traits early in a life history, can have profound negative effects later in that life history.

Farhad Manjoo's (2018) article "The Problem with Fixing WhatsApp? Human Nature Might Get in the Way" concerns the very human tendency to share information with others, and to trust others. Recounting a study of the spread of false news through WhatsApp, the author notes that the story of the transmission of the falsities "isn't of malicious and indiscriminate rumor-mongering. . . . It is, rather, a story of a few people who trusted other people, who in turn trusted others, each passing along what he or she considered important and necessary information to friends and colleagues." Manjoo then makes use of the human nature concept: "It's a story of human nature. And that's why, beyond learning to inhibit our natural tendency to share, it's hard to know what can be done about false news on WhatsApp – other than bracing yourself for more."

What Manjoo is identifying in human nature is our tendency to pass on information to others that we received from sources we trust. This can thus be understood as an observation about how traits are distributed over life histories. The antecedent trait is the reception of information, and the consequent is delivering the information to others. Manjoo argues that WhatsApp is just a means of triggering this entrenched pattern of behavior. Such a use of human nature thus beautifully fits the LTC account.

These examples show that at least some uses of the term "human nature" in the popular media assume a trait cluster account. While a concept of human nature should not be judged to stand or fall based on how well it fits with such uses, I feel it is an advantage of the LTC account that it helps us to understand references to human nature within and outside of academia. It shows that there is less disunity, less of a gap in understanding across diverse references to human nature.

6.5 Can Human Nature Guide Our Moral Behavior?

How might human nature tell us what is morally right or wrong? Can we derive moral norms from human nature? Or does human nature merely provide insights into how we are, not what we ought to be like? In an earlier paper (Ramsey 2012), I considered this problem in the context of *enhancement*, whether human nature can offer guidance in issues of human enhancement. Enhancement is a good test case for the moral implications of human nature. Is it morally sound to give children drugs that help them concentrate and learn better? Is it morally permissible for parents to be able to modify the genes of their embryos before they turn into children? Do such modifications go against human nature? If so, does making them mean that they are immoral? As I noted in my 2012 article, there are four important possibilities for the role of human nature:

1. Human nature could show us which traits are susceptible to enhancement.
2. Human nature could elucidate the risks or benefits of enhancement projects.
3. Human nature could enumerate the "natural" traits, providing us with a target for enhancement.
4. Human nature could provide us with new moral principles about what should and should not be enhanced.

Because the LTC account of human nature concerns how traits are distributed over possible life histories, knowing our nature includes knowing how consequent traits are dependent on antecedent traits. Knowledge of these trait relations can provide information on how interventions might modify consequent trait development. Thus, the LTC account of human nature can help to show us (1) which traits are susceptible to enhancement, and also (2) the risks and benefits of enhancement projects. But because the LTC account is not a trait bin account, (3) it does not sort traits into *natural* and *unnatural* bins, and thus, cannot enumerate the "natural" traits, providing us with a target for enhancement. Finally, while we can use moral premises in conjunction with human nature to derive new moral principles, (4) human nature by itself does not provide us with new moral principles about what should and should not be enhanced.

The LTC account of human nature, therefore, is not irrelevant to the moral dimensions of biomedical enhancement – or other forms of enhancement, such as education – but we should not look to human nature as an oracle for enlightening us as to which traits are natural and which are not. Human nature alone does not privilege some traits, does not indicate which traits should be avoided and which embraced. But by showing how traits are related to one another we can use this knowledge in conjunction with our moral compass to make decisions about which interventions we should avoid and which we should encourage.

This implication goes well beyond enhancement and applies generally to the relationship between the sciences and the morality of our actions. Human nature concerns the nexus of human traits and provides information about the counterfactual relations among our actions and reactions. What science teaches us about our nature thus has moral implications through the guidance we can achieve about these counterfactuals.

7 Conclusions

You observe a rushing river with churning rapids and white water and wonder whether this raucous violent state is the true nature of the river. If we travel back in time, we may find that the river used to be a small creek with calm, clear, flowing water. We may conclude from this that the true nature of the river, what is at its core, is a tranquil stream. We may make a similar inference with humans. As a proxy for our past, we could look to hunter-gatherers for a window into our nature. If we find them to be egalitarian, we may think that is our nature. If we find them to be murderous, we may think that is our nature. It is tempting to hold that our nature is hidden in our past or deep within ourselves. It is something beyond how we behave and think and feel. It is perhaps even occult, something divined through scripture or revelation.

My argument in this Element has been that human nature is not occult, that we wear our nature on our sleeve. To study human nature is to study humans here and now, our psychology and behavior and morphology. There is nothing to the nature of the river other than its form and behavior. Studying how the water flows – how it erodes the banks, how it speeds up in narrow channels, how it thunders at the base of cataracts – is studying the nature of the river. No sacred scripture is needed to tell of its nature, no prophets required to divine its hidden essence, no history to speak to what the river is at its core.

This view of human nature may be disappointing, like finding out that there really is no Santa Claus or Tooth Fairy. There is no occult power behind the arrival of presents; it is just the surreptitious action of parents or guardians. Similarly, there is no occult essence producing our thoughts and actions. It is merely the working of the machinery of life. As such, this view of human nature is a bit of a letdown. But the goal of this account of human nature is to seek the truth, not to spin exciting tales.

In *Fear and Trembling*, Kierkegaard wrote that "fools and young people chatter about everything being possible for a human being. However, this is a great misapprehension" (2006, 37). The LTC account is about articulating precisely how this is a misapprehension. It examines human lives, traces out what is possible and impossible, what is probable and improbable, which traits lead to or preclude others. By carving away what is impossible and by studying the articulations of what remains, we come to know our nature.

Want to know our nature? Then observe humans, their behavior, their artifacts. Read science, read fiction, listen to music. Stroll through museums and cast your eyes on sculptures and paintings and photographs. Go to sporting events, dinner parties, business meetings, classes, restaurants, bars. Sit on benches and walk through neighborhoods and go shopping and ride your bike. Eat a meal and have a conversation and get in a fight and have sex. Listen to human nature pulsing and reverberating within and around us, drink it in and ponder our place in the world.

References

Almécija, S., Hammond, A. S., Thompson, N. E. et al. (2021). Fossil apes and human evolution. *Science*, **372**(6542), eabb4363. https://doi.org/10.1126/sci ence.abb4363.

Bisbey, T. M., Reyes, D. L., Traylor, A. M., and Salas, E. (2019). Teams of psychologists helping teams: The evolution of the science of team training. *American Psychologist*, **74**(3), 278–89. https://doi.org/10.1037/amp000 0419.

Blake, P. R., McAuliffe, K., and Warneken, F. (2014). The developmental origins of fairness: The knowledge–behavior gap. *Trends in Cognitive Sciences*, **18**(11), 559–61. https://doi.org/10.1016/j.tics.2014.08.003.

Blake, P. R., McAuliffe, K., Corbit, J. et al. (2015). The ontogeny of fairness in seven societies. *Nature*, **528**, 258–61. https://doi.org/10.1038/nature15703.

Bouchard, T. J. (2013). The Wilson effect: The increase in heritability of IQ with age. *Twin Research and Human Genetics*, **16**(5), 923–30. https://doi.org/10 .1017/thg.2013.54.

Brosnan, S. F., and de Waal, F. B. M. (2003). Monkeys reject unequal pay. *Nature*, **425**, 297–9. https://doi.org/10.1038/nature01963.

Buller, D. J. (2005). *Adapting Minds: Evolutionary Psychology and the Persistent Quest for Human Nature*. Cambridge, MA: MIT Press.

Cashdan, E. (2013). What is a human universal? Human behavioral ecology and human nature. In S. Downes and E. Machery, eds., *Arguing about Human Nature: Contemporary Debates*. New York: Routledge Press, pp. 71–80.

Darwin, C. (1859). *On the Origin of Species by Means of Natural Selection*. London: Murray.

De Waal, F. B. M. (2005). *Our Inner Ape: A Leading Primatologist Explains Why We Are Who We Are*. New York: Riverhead Books.

Doris, J. M. (2002). *Lack of Character: Personality and Moral Behavior*. Cambridge: Cambridge University Press.

Dubreuil, D., Gentile, M. S., and Visalberghi, E. (2006). Are capuchin monkeys (*Cebus apella*) inequity averse? *Proceedings of the Royal Society B: Biological Sciences*, **273**, 1223–8. https://doi.org/10.1098/rspb.2005.3433.

Dupré, J. (2018). Human nature: A process perspective. In E. Hannon and T. Lewens, eds., *Why We Disagree about Human Nature*. Oxford: Oxford University Press, pp. 92–107.

Eiberg, H., Troelsen, J., Nielsen, M. et al. (2008). Blue eye color in humans may be caused by a perfectly associated founder mutation in a regulatory element located within the HERC2 gene inhibiting OCA2 expression. *Human Genetics*, **123**, 177–87. https://doi.org/10.1007/s00439-007-0460-x.

Ellis, S., Franks, D. W., Nattrass, S. et al. (2018). Analyses of ovarian activity reveal repeated evolution of post-reproductive lifespans in toothed whales. *Scientific Reports*, **8**(1), 1–10. https://doi.org/10.1038/s41598-018-31047-8.

Evershed, R. P., Payne, S., Sherratt, A. G. et al. (2008). Earliest date for milk use in the near East and Southeastern Europe linked to cattle herding. *Nature*, **455**, 528–31. https://doi.org/10.1038/nature07180.

Evershed, R. P., Davey Smith, G., Roffet-Salque, M. et al. (2022). Dairying, diseases and the evolution of lactase persistence in Europe. *Nature*, **608** (7922), 336–45.

Freud, S. (1962). *Civilization and Its Discontents*. New York: W. W. Norton.

Geraci, A., and Surian, L. (2011). The developmental roots of fairness: Infants' reactions to equal and unequal distributions of resources. *Developmental Science*, **14**(5), 1012–20. https://doi.org/10.1111/j.1467-7687.2011.01048.x.

Ghiselin, M. T. (1974). *The Economy of Nature and the Evolution of Sex*. Berkeley: University of California Press.

Ghiselin, M. T. (1997). *Metaphysics and the Origins of Species*. Albany: SUNY Press.

Glazko, G. V., and Masatoshi, N. (2003). Estimation of divergence times for major lineages of primate species. *Molecular Biology and Evolution*, **20**, 424–34. https://doi.org/10.1093/molbev/msg050.

Greene, R. (2018). *The Laws of Human Nature*. New York: Viking.

Griffiths, P. (1999). Squaring the circle: Natural kinds with historical essences. In R. A. Wilson, ed., *Species: New Interdisciplinary Essays*. Cambridge, MA: MIT Press, pp. 209–28.

Griffiths, P. (2009). Reconstructing Human Nature. *Arts: The Journal of the Sydney University Arts Association*, **31**, 30–57.

Griffiths, P. (2011). Our plastic nature. In S. B. Gissis and E. Jablonka, eds., *Transformations of Lamarckism: From Subtle Fluids to Molecular Biology*. Cambridge, MA: MIT Press, pp. 319–30.

Griffiths, P., Machery, E., and Linquist, S. (2009). The vernacular concept of innateness. *Mind & Language*, **24**(5), 605–30. https://doi.org/10.1111/j.1468-0017.2009.01376.x.

Gupta, A. P., and Lewontin, R. C. (1982). A study of reaction norms in natural populations of *Drosophila pseudoobscura*. *Evolution*, **36**(5), 934–48. https://doi.org/10.2307/2408073.

Hacker, P. M. S. (2021). *The Moral Powers: A Study of Human Nature*. New York: John Wiley.

Hannon, E., and Lewens, T. (2008). *Why We Disagree about Human Nature.* Oxford: Oxford University Press.

Hartmann, W. E., Wendt, D. C., Burrage, R. L., Pomerville, A., and Gone, J. P. (2019). American Indian historical trauma: Anticolonial prescriptions for healing, resilience, and survivance. *American Psychologist*, **74**(1), 6–19. https://doi.org/10.1037/amp0000326.

Hoffman, L., and Oppenheim, L. (2019). Three identical strangers and the twinning reaction – clarifying history and lessons for today from Peter Neubauer's twins study. *JAMA*, **322**, 10–12. https://doi.org/10.1001/jama .2019.8152.

Hull, D. L. (1986). On human nature. *PSA: Proceedings of the Biennial Meeting of the Philosophy of Science Association*, **2**, 3–13. https://doi.org/10.1086/ psaprocbienmeetp.1986.2.192787.

Hume, D. (1731). *A Treatise of Human Nature.* Oxford: Clarendon Press.

Hyde, J. S., Bigler, R. S., Joel, D., Tate, C. C., and van Anders, S. M. (2019). The future of sex and gender in psychology: Five challenges to the gender binary. *American Psychologist*, **74**(2), 171–9. https://doi.org/10.1037/amp0000307.

Ingold, T. (2019). Review of *Why We Disagree about Human Nature. Notre Dame Philosophical Reviews.* https://ndpr.nd.edu/news/why-we-disagree-about-human-nature/.

Ingram, C. J., Mulcare, C. A., Itan, Y., Thomas, M. G., and Swallow, D. M. (2009). Lactose digestion and the evolutionary genetics of lactase persistence. *Human Genetics*, **124**, 579–91. https://doi.org/10.1007/s00439-008-0593-6.

Kellogg, W. N., and Kellogg, L. A. (1933). *The Ape and the Child: A Study of Environmental Influence upon Early Behavior.* New York: Whittlesey House.

Kierkegaard, S. (2006). *Kierkegaard: Fear and Trembling.* Edited by C. S. Evans and S. Walsh. Cambridge: Cambridge University Press.

Kippenhahn, R., Weigert, A., and Weiss, A. (1990). *Stellar Structure and Evolution.* Berlin: Springer-Verlag.

Kronfeldner, M. (2018). *What's Left of Human Nature? A Post-essentialist, Pluralist, and Interactive Account of a Contested Concept.* Cambridge, MA: MIT press.

Kuhn, T. S. (1962). *The Structure of Scientific Revolutions.* Chicago: University of Chicago Press.

LaPorte, J. (2004). *Natural Kinds and Conceptual Change.* Cambridge: Cambridge University Press.

Leonardi, M., Gerbault, P., Thomas, M. G., and Burger, J. (2012). The evolution of lactase persistence in Europe: A synthesis of archaeological and genetic evidence. *International Dairy Journal*, **22**(2), 88–97. https://doi.org/10.1016/ j.idairyj.2011.10.010.

Lipkis, E. (2017). Blame human nature, not guns. *New York Times*, November 9.

Machery, E. (2008). A plea for human nature. *Philosophical Psychology*, **21**(3), 321–9. https://doi.org/10.1080/09515080802170119.

Machery, E. (2016). Human nature. In D. L. Smith, ed., *How Biology Shapes Philosophy: New Foundations for Naturalism*. Cambridge: Cambridge University Press, pp. 204–26.

Manjoo, F. (2018). The problem with fixing WhatsApp? Human nature might get in the way. *New York Times*, October 24.

McAuliffe, K., Chang, L. W., Leimgruber, K. L. et al. (2015). Capuchin monkeys, *Cebus apella*, show no evidence for inequity aversion in a costly choice task. *Animal Behaviour*, **103**, 65–74. https://doi.org/10.1016/j.anbehav.2015.02.014.

McAuliffe, K., Blake, P. R., and Warneken, F. (2017). Do kids have a fundamental sense of fairness? Experiments show that this quality often emerges by the age of 12 months. *Scientific America* blog, August 23. https://blogs.scientificamerican.com/observations/do-kids-have-a-fundamental-sense-of-fairness/?print=true.

McCormack, W. (2018). Records from controversial twin study sealed at Yale until 2065. *Yale Daily News*, October 1. https://yaledailynews.com/blog/2018/10/01/records-from-controversial-twin-study-sealed-at-yale-until-2065/.

Mischel, W., Shoda, Y., and Mendoza-Denton, R. (2002). Situation-behavior profiles as a locus of consistency in personality. *Current Directions in Psychological Science*, **11**(2), 50–4. https://doi.org/10.1111/1467-8721.00166.

Owen, R. (1843). *Lectures on the Comparative Anatomy and Physiology of the Invertebrate Animals: Delivered at the Royal College of Surgeons, in 1843*. London: Longman.

Pedroso, M. (2012). Essentialism, history, and biological taxa. *Studies in History and Philosophy of Science Part C: Studies in History and Philosophy of Biological and Biomedical Sciences*, **43**(1), 182–90. https://doi.org/10.1016/j.shpsc.2011.10.019.

Peirson, B. R. E. (2012). Richard Woltereck's concept of Reaktionsnorm. *Embryo Project Encyclopedia* (2012-09-06). ISSN: 1940-5030. http://embryo.asu.edu/handle/10776/3940.

Pinker, S. (2002). *The Blank Slate: The Modern Denial of Human Nature*. New York: Viking.

Quain, J. R. (2016). Makers of Self-Driving Cars Ask What to Do with Human Nature. *New York Times*, July 7.

Ramsey, G. (2012). How human nature can inform human enhancement: A commentary on Tim Lewens's "Human nature: The very idea." *Philosophy and Technology*, **25**, 479–83. https://doi.org/10.1007/s13347-012-0087-2.

Ramsey, G. (2013). Human nature in a post-essentialist world. *Philosophy of Science*, **80**, 983–93. https://doi.org/10.1086/673902.

Ramsey, G. (2017). What is human nature for? In A. Fuentes and A. Visala, eds., *Verbs, Bones, and Brains: Interdisciplinary Perspectives on Human Nature*. Notre Dame: University of Notre Dame Press, pp. 217–30.

Ramsey, G. (2018). Trait bin and trait cluster accounts of human nature. In E. Hannon and T. Lewens, eds., *Why We Disagree about Human Nature*. Oxford: Oxford University Press, pp. 40–57.

Ramsey, G., and Peterson, A. S. (2012). Sameness in biology. *Philosophy of Science*, **79**(2), 255–75. https://doi.org/10.1086/664744.

Rochat, P., Dias, M. D., Liping, G. et al. (2009). Fairness in distributive justice by 3- and 5-year-olds across seven cultures. *Journal of Cross-Cultural Psychology*, **40**(3), 416–42. https://doi.org/10.1177/0022022109332844.

Shoda, Y., LeeTiernan, S., and Mischel, W. (2002). Personality as a dynamical system: Emergence of stability and distinctiveness from intra- and interpersonal interactions. *Personality and Social Psychology Review*, **6**(4), 316–25. https://doi.org/10.1207/S15327957PSPR0604_06.

Sloane, S., Baillargeon, R., and Premack, D. (2012). Do infants have a sense of fairness? *Psychological Science*, **23**(2), 196–204. https://doi.org/10.1177/0956797611422072.

Sommerville, J. A., Schmidt, M. F. H., Yun, J. E., and Burns, M. (2013). The development of fairness expectations and prosocial behavior in the second year of life. *Infancy*, **18**, 40–66. https://doi.org/10.1111/j.1532-7078.2012.00129.x.

Sterelny, K. (2018). Skeptical reflections on human nature. In E. Hannon and T. Lewens, eds., *Why We Disagree about Human Nature*. Oxford: Oxford University Press, pp. 108–26.

Stevenson, L. F., Haberman, D. L., Wright, P. M., and Witt, C. (2017). *Thirteen Theories of Human Nature*. Oxford: Oxford University Press.

Stotz, K., and Griffiths, P. (2018). A developmental systems account of human nature. In E. Hannon and T. Lewens, eds., *Why We Disagree about Human Nature*. Oxford: Oxford University Press, pp. 58–75.

Swallow, D. M. (2003). Genetics of lactase persistence and lactose intolerance. *Annual Review of Genetics*, **37**(1), 197–219. https://doi.org/10.1146/annurev.genet.37.110801.143820.

Acknowledgments

I thank Maya Parson and Michael Ruse for generously reading and commenting on multiple drafts of this Element. And I thank Andreas De Block, James DiFrisco, Alex Gordillo-Garcia, Paul Griffiths, Andra Meneganzin, Brian McLoon, and the anonymous reviewers for their valuable feedback. Finally, I thank the Research Foundation Flanders (FWO) for their support – in particular, for grant G070122 N.

Cambridge Elements ≡

Philosophy of Biology

Grant Ramsey

KU Leuven, Belgium

Grant Ramsey is a BOFZAP research professor at the Institute of Philosophy, KU Leuven, Belgium. His work centers on philosophical problems at the foundation of evolutionary biology. He has been awarded the Popper Prize twice for his work in this area. He also publishes in the philosophy of animal behavior, human nature and the moral emotions. He runs the Ramsey Lab (theramseylab.org), a highly collaborative research group focused on issues in the philosophy of the life sciences.

Michael Ruse

Florida State University

Michael Ruse is the Lucyle T. Werkmeister Professor of Philosophy and the Director of the Program in the History and Philosophy of Science at Florida State University. He is Professor Emeritus at the University of Guelph, in Ontario, Canada. He is a former Guggenheim fellow and Gifford lecturer. He is the author or editor of over sixty books, most recently *Darwinism as Religion: What Literature Tells Us about Evolution*; *On Purpose*; *The Problem of War: Darwinism, Christianity, and their Battle to Understand Human Conflict*; and *A Meaning to Life*.

About the Series

This Cambridge Elements series provides concise and structured introductions to all of the central topics in the philosophy of biology. Contributors to the series are cutting-edge researchers who offer balanced, comprehensive coverage of multiple perspectives, while also developing new ideas and arguments from a unique viewpoint.

Cambridge Elements ☰

Philosophy of Biology

Elements in the Series

A full series listing is available at www.cambridge.org/EPBY

Printed in the United States
by Baker & Taylor Publisher Services